THE OUTER LANDS

THE OUTER LANDS Revised Edition

A Natural History Guide to Cape Cod, Martha's Vineyard, Nantucket, Block Island, and Long Island

DOROTHY STERLING

ILLUSTRATED BY WINIFRED LUBELL

Foreword to the revised edition
by Robert Finch

W · W · NORTON & COMPANY

New York · London

W. W. Norton & Company, Inc., 500 Fifth Avenue, New York, N.Y. 10110
W. W. Norton & Company Ltd., 37 Great Russell Street, London WC1B 3NU

Printed in The United States of America
Revised Edition

Library of Congress Cataloging in Publication Data
Sterling, Dorothy
The outer lands.
1. Seashore biology—Massachusetts—Cape Cod region. 2. Seashore biology—New York (State)—Long Island. 3. Natural history—Massachusetts—Cape Cod region. 4. Natural history—New York (State)—Long Island. I. Lubell, Winifred. II. Title.
QH105.M4S73 1978 574.9744'9 78-2339

5 6 7 8 9 0

ISBN 0-393-06438-7
ISBN 0-393-06441-7 pbk.

CONTENTS

FOREWORD

One day last summer I overheard a family of tourists as they viewed the exhibits in the Cape Cod Museum of Natural History, where I work. The two young boys raced omnivorously from one display to another, gawking at the strange goose-necked barnacles in the salt-water aquarium, handling the necklace-like whelk egg cases, mugging at the frogs and turtles in the aquatics building. Their mother, whose primary concern seemed to be to keep the boys from getting very near anything, turned at one point to her husband and remarked: "I don't see what's so special about this place. We can see these things down on the beach!"

Exactly. We live in an age that devalues the familiar and the near at hand, that has taught us to seek novelty rather than reality. Conditioned by electronic media to instant and effortless gratification, we are impatient with a nature that demands deliberate and active response from our mind and senses. We would rather pay for fictional "Close Encounters" with celluloid aliens than seek actual encounters with a jellyfish or a mole crab, though I think the latter type more genuinely stretches our imaginations.

In an age of global travel, our need for local identification has never been greater. Today's expressways have given us increased speed but less perception, enabling larger numbers of us to go more and more places to see less and less. The glacial outlands described in this book represent some of the oldest settled portions of our country. They front the northeast

Megalopolis and are within a day's drive to over seventy million people. Each summer countless feet mark their beaches, and new cottages, motels, and year-round houses continue to spring up at a rapid rate.

Yet for all their human traffic, they remain largely an undiscovered country. We have lost that recognition of the familiar that was the common birthright of the old Cape Codders and Islanders, a recognition that came from their intimate and daily interaction with the tides, shores, and marshes that gave them a living. We, on the other hand, having tricked ourselves for so long into believing that we are independent of the life around us, use such places narrowly, oversimplifying them, and so failing to see their diversity and complexity.

This is what makes a book like *The Outer Lands* so valuable. By making us aware of other worlds of equal stature, complexity, beauty, and wonder that share our own territory, it can help to rescue us from that terrible, isolating human provincialism that our modern mobility has, ironically, bred. It is a "guide" in the best sense, for it not only informs but heightens our expectations about what we might see. Its pages communicate the author's and the artist's own affection and delight, born of long acquaintance, stimulating in us an appetite for rediscovery and providing us with a rich sense of unexplored possibilities in an all-too-familiar landscape.

It is, after all, our own expectations that primarily determine what we find or do not find in the natural world. If these are cheap and easy, then we can count on being disappointed and will no doubt drift gratefully back to our television sets with a new gratitude and appreciation for the "gifts of civilization."

But if, using such guides as this, we can acquaint ourselves with a place without thinking that we know it; if we can learn enough facts to gain entrance, but not so much as to draw a curtain of certainty over what we might encounter there; if we have cocked our ears, but do not require what we hear to have a human sound; and, finally, if we seek confrontation rather than confirmation, so that even our chance meeting with a kingfisher or a patch of sea lavender risks our most cherished conceptions of the world—then we might see somethting.

Dorothy Sterling talks of these wind-swept, sea-girt, sandy shores as "the remnants of an ancient coastline." They are that, old lands that can give us a new appreciation of the depth of our geologic history. But they are also forever new, forever changing places where the ongoing creation of this planet, in all its marvelous diversity, is somehow peculiarly evident, immediate, and accessible on a human scale. New worlds are ours for the asking here; *The Outer Lands*, with beautiful illustrations by Winifred Lubell, offers its readers an eloquent invitation to discovery.

Robert Finch

THE OUTER LANDS

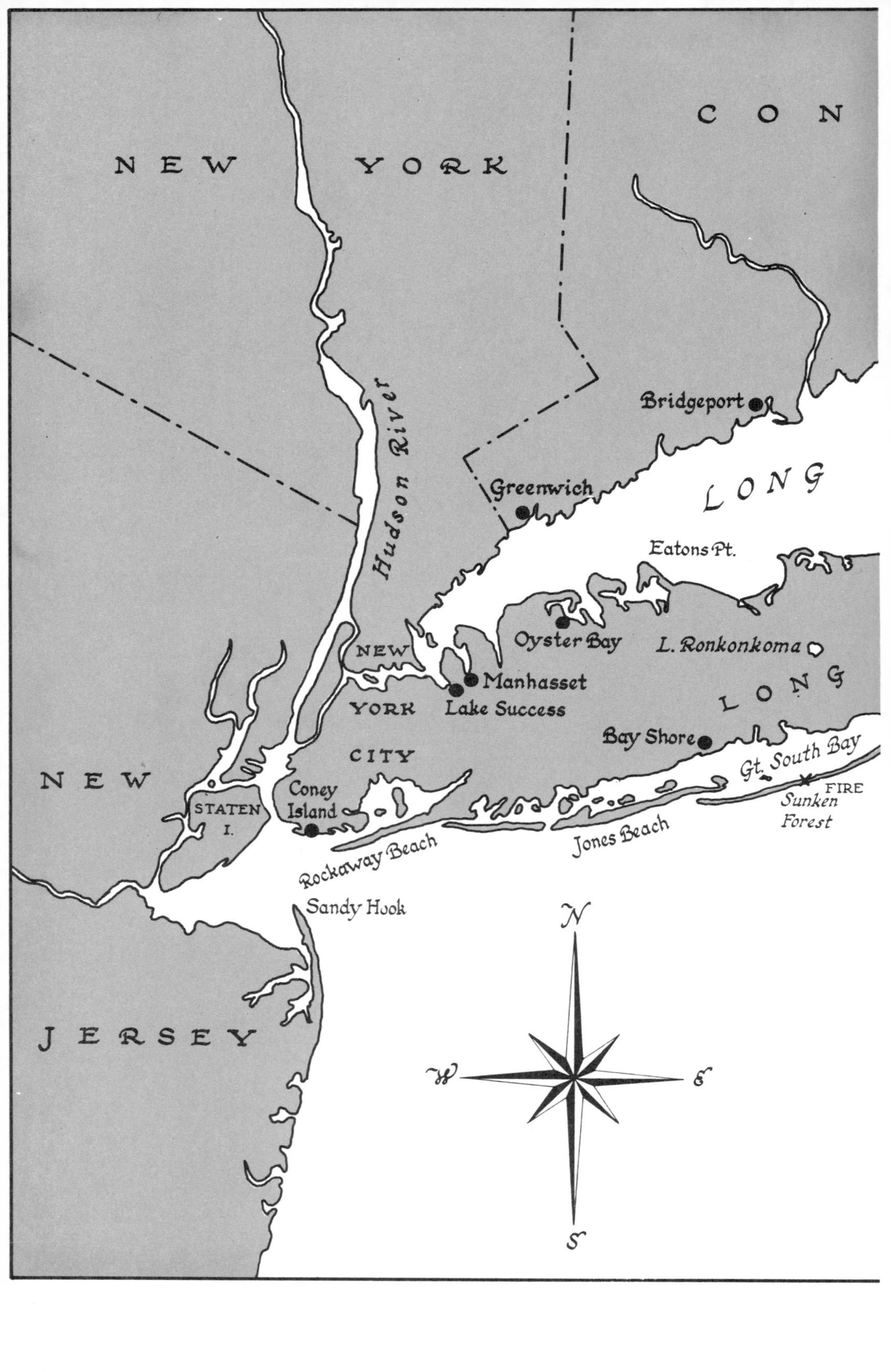

NEW YORK
CON
Hudson River
Bridgeport
Greenwich
LONG
Eatons Pt.
Oyster Bay
L. Ronkonkoma
NEW
YORK
CITY
Manhasset
Lake Success
LONG
Bay Shore
Gt. South Bay
NEW
Coney
Island
STATEN
I.
FIRE
Sunken
Forest
Jones Beach
Rockaway Beach
Sandy Hook
JERSEY
N
W
E
S

NECTICUT
RHODE ISLAND
Mystic
New London
New Haven
Pt. Judith
FISHERS I.
Sandy Pt.
ISLAND SOUND
Gt. Salt Pond
Orient Point
BLOCK I.
GARDINERS I.
Gardiners Bay
Montauk Point
Peconic Bay
East Hampton
Riverhead
ISLAND
Southampton
Gt. South Beach
ISLAND
ATLANTIC OCEAN
LONG ISLAND AREA
0
5
10
15
20
25
Nautical Miles
deFontaine

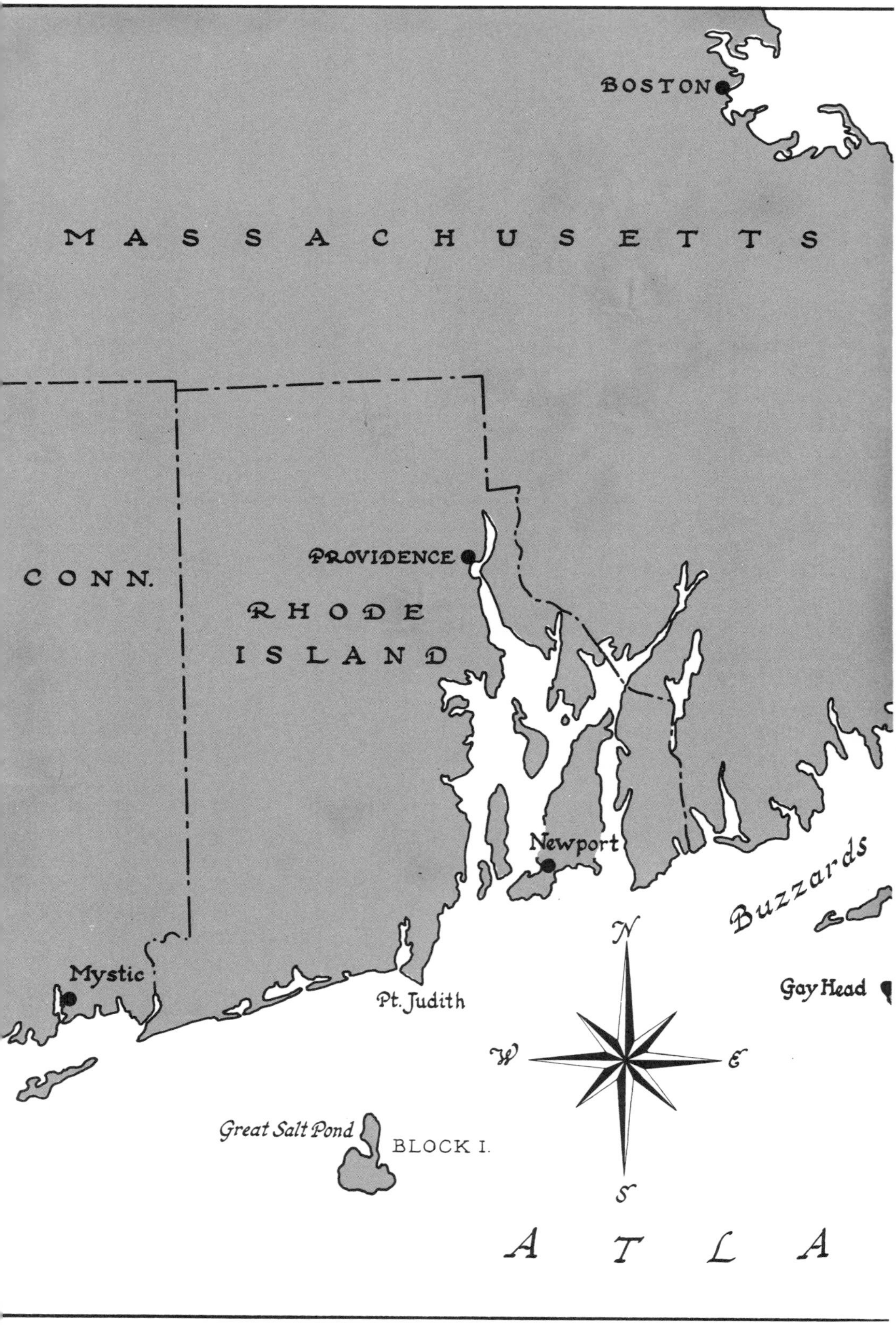

BOSTON

MASSACHUSETTS

PROVIDENCE

CONN.

RHODE ISLAND

Newport

Buzzards

Mystic

Pt. Judith

Gay Head

N

W

E

S

Great Salt Pond

BLOCK I.

ATLA

CAPE COD AREA
Provincetown
Truro
Wellfleet
Plymouth
Cape Cod Bay
Eastham
Nauset Beach
Orleans
Sagamore
Cape Cod Canal
Sandwich
Barnstable
CAPE COD
Chatham
Bay
Cotuit
Monomoy
Falmouth
Nantucket Sound
Woods Hole
Cape Poge
Edgartown
Coatue Beach
MARTHA'S VINEYARD
Tuckernuck I.
Nantucket
NO MANS LAND
NANTUCKET
N T I C
O C E A N
0 5 10 15 20 25
Nautical Miles
deFontaine

PREFACE

No continent is entire of itself. Between the Atlantic Ocean and the rocky rim of New England lies a chain of islands and a peninsula. Long Island points northeast across open stretches of water to Block Island, Martha's Vineyard, Nantucket and the flexed arm of Cape Cod. Isolated by the awesome processes of the earth's evolution, these sandy, windswept outlands are the remnants of an ancient coastline. Although they have a shared geology and a common natural history, man, with little concern for this kinship, has parceled them out to three states. Their residents vote for different congressmen, pay different taxes, obey different traffic laws. Even their diet and speech are different. Cape Cod's famous clam chowder is made from quahogs and milk, gently boiled, while Long Islanders prepare their chowder with hard-shell clams and stewed tomatoes.

Yet anyone who has walked along the Cape and island beaches, collected shells, dug for clams, fished for stripers, knows that their man-designated differences are less important than their natural similarities. Waves and wind, sun and fog have shaped the sandy substance left behind by Ice Age glaciers to create a landscape strikingly different from that of the mainland, while the lands' proximity to the sea has given them a unique plant and animal life. There are broad beaches of white sand, shimmering dunes and green salt meadows. There are quiet ponds, rolling moors and twisted pines. Even the quahog of Cape Cod and Long Island's hard-shell clam are the same organism—and only the argument between milk and tomatoes remains to be settled.

This book is a survey of the natural history of these Outer Lands—their dynamic landscape and their distinctive animal and plant life. Confining itself to the ecological communities along the shore and within range of sea wind and salt spray, it makes no attempt to report on inland areas whose birds, flowers, trees are similar to those on the mainland. In addition to identifying animals and plants, the book attempts to tell something of their life processes and interlocking relationships.

I BIRTH OF THE OUTER LANDS

The Big Ice

The glacier grew from year to year. During century-long winters cold winds whistled and snow piled on top of snow. The mountain of blue ice thickened until the pressure of its own weight set it moving. Grinding down from Labrador to cover more than half of North America, the glacier put the finishing touches on our continent. Nowhere is its imprint more clearly seen than on the Outer Lands—the sandy hills and scrub-covered moors that string out along the shores of the Atlantic from Long Island to Cape Cod.

Millions of years before the Great Ice, these lands were part of an ancient coastal plain where forests grew and dinosaurs roamed. Time and again invading seas drowned the forests and left them covered with mud and clay. The dinosaurs disappeared and eons later camels and horses grazed on the marshy plains while giant sharks swam in the waters offshore.

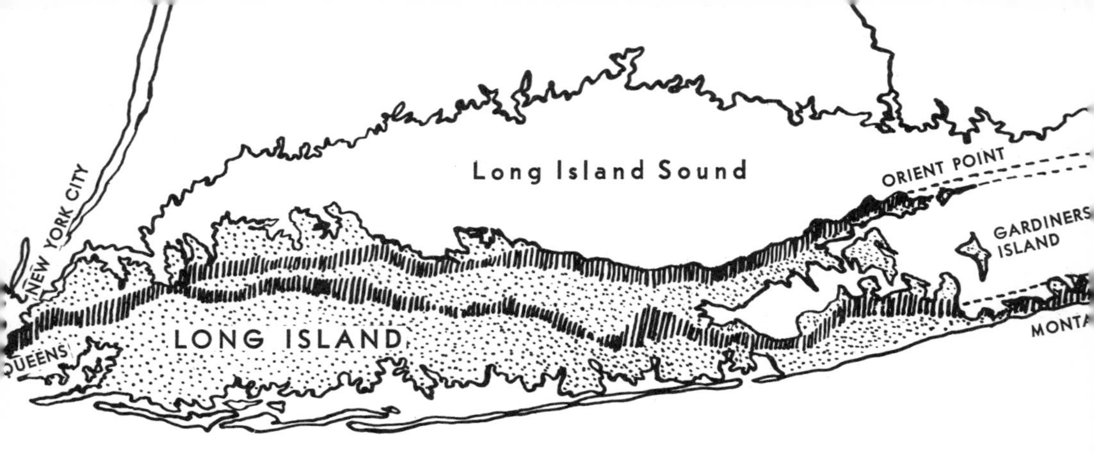

The Outer Lands

MORAINES

OUTWASH PLAINS

Some thirty thousand years ago—only yesterday as geologic time is figured—the glacier began its fateful march. As it advanced through New England, the mile-thick ice sheet sliced off mountain peaks, flattened hills and widened valleys. Like a giant steam shovel it dug into the rocky core of the continent. Prying boulders loose from their beds, it ground them in its mill of moving ice.

The Hills

When the glacier reached the soft margins of the coastal plain it scooped out the basins of bays and sounds and pushed on toward the sea. Halted by currents of warm air, it dumped the load of rocks and pebbles it had carried from the mainland. These glacial dump heaps, torn from once-solid bedrock, formed ridges hundreds of feet high. Known as moraines, they are the framework of the Outer Lands.

The southernmost advance of the glacier can be seen

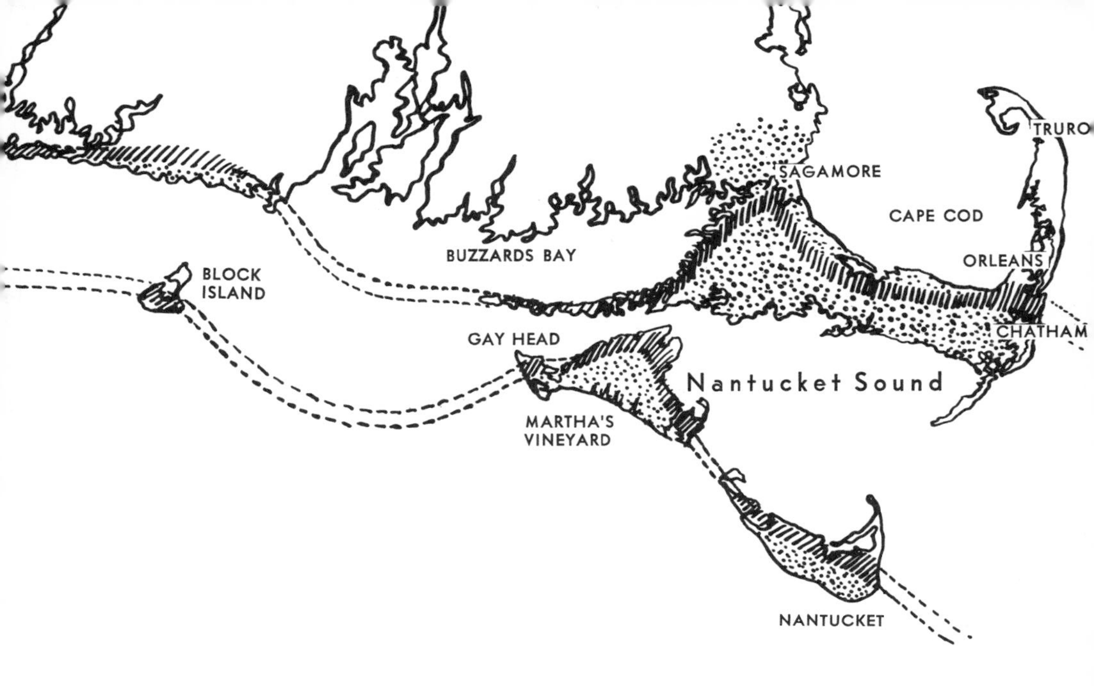

today on Long Island in the double row of hills that stretches from Queens to Orient Point and Montauk. Off Montauk, the moraine has been covered by rising seas, but it reappears again in the hills of Block Island, Martha's Vineyard and Nantucket. On Cape Cod the scalloped edges of the glacier left separate morainal hills. One runs north and south along Buzzards Bay, the other follows the shore of Cape Cod Bay from Sagamore to Orleans, then turns south to Chatham.

A hundred centuries of trees and grasses have clothed these uplands with a top dressing of soil, but minutes of digging will uncover the stones and pebbles of which they are built—the gravelly material that geologists call till. Sometimes, when the grinding ice failed to digest its rocky burden, it dropped boulders. These boulders are scattered across the Outer Lands —"like nuts in a frosting," one writer said—but they are most prominent along Long Island's north shore and in the Falmouth-Woods Hole area of Cape Cod.

Enos Rock in Eastham, shadowed by a grove of

Gay Head Cliffs, Martha's Vineyard

cedars, is the largest boulder on Cape Cod. Whether it is bigger than the boulder that dominates a parking lot in Manhasset, as local pride claims, won't be known until someone digs them out. Since each is estimated to weigh thousands of pounds, this is not likely to happen soon.

In a few places, the hills of the Outer Lands are not built of glacial till. There were long periods of the Ice Age when the glacier retreated and the waters of the ocean rose. During one of these times beds of blue clay formed in the sea. Known as Gardiners Clay because it makes up most of the subsoil of Gardiners Island, it can be traced from eastern Long Island to Block Island, Martha's Vineyard and Cape Cod. Thick deposits of this clay—which Indians and early settlers used for pottery—are most easily seen at Mohegan Bluffs on

Block Island, Nashaquitsa Cliffs on the Vineyard and the Clay Pounds of Truro, just north of Cape Cod Light.

The cliffs at Gay Head on Martha's Vineyard are also an exception. Here the highlands record the million centuries before the Ice Age. Each brightly colored twisted layer of the cliffs tells the story of a different time in the history of the ancient coastal plain. Buried forests, perhaps the very same ones that gave shelter to dinosaurs, form a layer of black at the base of the cliffs. Clays, stained yellow and red by iron and minerals, and a layer of greensand (which despite its name looks red after it is exposed to air) contain fossils of clam shells and crabs. In the gravelly layers near the top, whale bones and sharks' teeth are common and bits of the skeleton of a camel and wild horse have been found. The town of Gay Head prohibits digging in the crumbling cliffs. However, sharp-eyed visitors who walk along the beach below can sometimes pick up fossils that have been washed out by rains.

Plains and Ponds

Whenever a warming sun shone on the great ice sheet, streams of water flowed from its surfaces. Swift-moving in summer, sluggish in winter, these rivers of meltwater cascaded down the face of the hills. Carrying fragments of rock and gravel, they built broad fan-shaped plains that sloped from the bottom of the hills to the ocean floor. Today these outwash plains are the potato fields and truck farms of Long Island, the

Barrier Beaches

moors of Nantucket, the Vineyard's Great Plains and Cape Cod's grassy "South Sea."

The glacier pulled back faster and faster, but here and there stubborn chunks of ice remained. Rushing streams detoured around these icy islands or, flowing over them, coated them with gravel. In time, as the ice blocks melted, they left rounded pockets in the sandy ground. Some of these kettle holes, as they are called, were only shallow bowls. Others were a hundred feet deep and a mile wide. Lying below the water table, they filled up with ground water to become ponds.

Lake Success and Lake Ronkonkoma on Long Island are kettle-hole ponds. So are Old House Pond and Fresh Pond on Martha's Vineyard and Washing and Gibbs Pond on Nantucket. Cape Cod has almost five hundred kettle-hole ponds—enough to drown every gossip in, Cape folks used to say. The Vineyard's Great Ponds and Nantucket's Hummock and Long ponds have a different origin, however. These narrow ponds were once meltwater streams that flowed across the outwash plains to the sea. Long after the glacier departed, the ocean walled them off with dams of sand. Block Island has many small kettle-hole ponds, but Great Salt Pond was originally a sea passage which

cut the island in two. Fenced in by sandbars in postglacial times, its western shore has been opened to the sea again to provide the islanders with a sheltered harbor.

The Sea's Work

If you had been able to fly over the Outer Lands at the end of the Ice Age you would have seen barren gravel banks. The hills and plains were there, dusted with boulders and pockmarked by kettle holes. But there were no beaches, no stretches of green marsh, no dunes or woods. Cape Cod, wider than it is today, ended abruptly at the Truro highlands. Long Island's south shore fronted on the Atlantic Ocean without the protective barrier of Fire Island or Jones Beach.

Seeds left by the glacier or dropped by birds sprouted in the gravelly soil. Animals traveled over from the mainland, following land bridges that no longer exist. And the sea began its mighty work.

Fed by melting ice, the waters of the ocean rose rapidly. They filled the shallow bays and sounds that the glacier had dug, flooded low-lying plains and began a steady, relentless attack on the hills. The bedrock

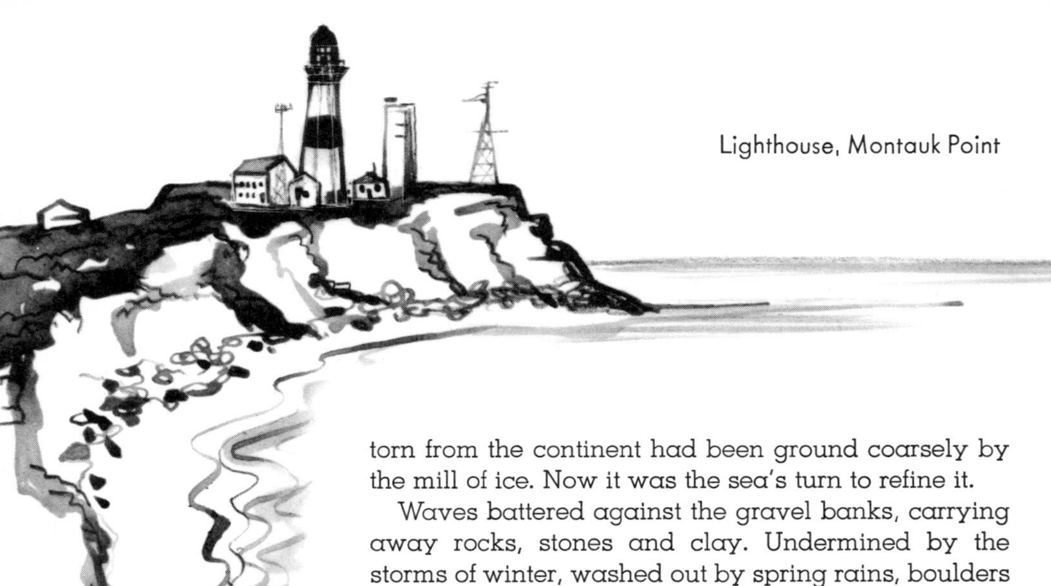

Lighthouse, Montauk Point

torn from the continent had been ground coarsely by the mill of ice. Now it was the sea's turn to refine it.

Waves battered against the gravel banks, carrying away rocks, stones and clay. Undermined by the storms of winter, washed out by spring rains, boulders tumbled from the bluffs. A million times, and a million times a million, the smaller fragments of the glacial till somersaulted along the ocean floor. Pulled by winds and pushed by currents, they were swept out to sea and tossed back to shore. Rocks rubbed against each other. Stones clattered over the surface of the boulders at high tide. Filing, scraping, polishing rough edges, the endless motion of the waves turned the coarse gravel into smooth pebbles and the pebbles into sand.

Over the centuries, the restless ocean laid down a broad border of sand around the Outer Lands. It shaped harbors and bays, carved inlets—and then proceeded to rearrange them. Borrowing newly made sand from the beaches, the waves built bars and shoals offshore. The bars became spits and the spits, peninsulas. The shoals grew into islands and the islands joined to form long barrier beaches. A barrier beach is exactly what its name implies. A narrow strip of sand running parallel to the shore, it acts as a barrier to the ocean. Protecting the land in front of it, it transforms open harbors into sheltered bays.

A glance at a map of the Outer Lands will show the work of the sea. On the Cape, glacial debris from the cliffs and sand from the beaches built the hooked arm of Provincetown, much of Nauset Beach and all of the sometime-island, sometime-peninsula of Monomoy. Nantucket's Coatue Beach is recent. So is the Vineyard's South Beach and Cape Poge. Along Long Island's south shore from Coney Island to Southampton the waves have thrown up almost a hundred miles of barrier beaches. Fire Island, Rockaway Beach, Jones Beach—these are all new lands formed since the Ice Age.

Like the Lord, the sea giveth and it taketh away. It builds and it destroys. Local histories are filled with accounts of lost islands and peninsulas. Nauset Island off Eastham which may—or may not—have been visited by Leif Ericson and which certainly was mapped by the French explorer Champlain, today lies under many feet of water. Billingsgate Island where Wellfleet farmers once pastured horses is now a sandbar visible only at low tide. Nantucket has lost two islands near Tuckernuck Island and Block Islanders recall that two centuries ago their forebears picked beach plums on a peninsula beyond Sandy Point.

Records of the lighthouses built along the coast show that the work of the sea is a continuing process. On the advice of no less an engineer than George Washington, Montauk Point Light, erected in 1797, was placed 200 feet back from the sea. There, he predicted, it would stand for 200 years. And there it has stood while waves hammered at the cliff in front of it. Today, after 170 years, less than 40 feet of land remains between the base of the lighthouse and the cliff's edge.

Apparently, President Washington was not consulted when Cape Cod Light was built a year later. Moved back once, it will soon be ready for a second retreat. Storms eat away the bluffs in this area at a rate that averages almost four feet a year. Lighthouses on Nantucket and the Vineyard have met similar fates. In exposed places like Cape Poge the wind-driven surf sometimes carries off ten feet of land in winter.

Wind and Weather

The wind plays its own part in the endless shaping and reshaping of the shores. Each sector of the Outer Lands boasts that it is the windiest spot on the North Atlantic. Although the Weather Bureau refuses to take sides it has recorded 91-mile-per-hour winds at its stations on Block Island and Nantucket—windy enough for any person's taste.

In summer the prevailing wind comes from the southwest. In winter it drives across the open ocean from the northeast with terrific force. Blasting the face of the cliffs, it starts avalanches of gravel that slide to the water's edge. While feeding the sea its daily diet of stones the wind also picks up sand grains and heaps them into dunes. And the dunes move too, creeping inland to build sand hills and valleys of their own.

Together the busy wind and the ceaseless ocean have tempered the climate of the Outer Lands. Oceans have a narrower range of temperatures than continents do. Their waters warm slowly in the spring and cool off slowly in the fall. Because the Cape and islands are almost completely surrounded by water, they are cooler in summer and warmer in winter than the mainland. In addition, the temperature of their coastal waters is influenced by two distinctly different ocean currents. The Labrador Current flows down the New England coast, bringing chilled Arctic waters to the beaches of Maine and Massachusetts. At the same time the Gulf Stream travels up from the south, swinging east to cross the Atlantic in the neighborhood of Nantucket. The near-meeting of the two currents, each with a different population of cold- or warm-water animals, gives the Outer Lands an unusually rich assortment of marine life.

5000 A.D.

Anyone who has listened to the pounding of the surf during a winter storm knows that the work of wind and waves still continues. People are now trying to slow down these powerful forces by building breakwaters and planting grass on the shifting dunes. Despite these efforts, some scientists believe that the Outer Lands are doomed. Over the long pull their real enemy may be

Enos Rock, Cape Cod

the very glacier that built them. If we are entering a period of warmer weather, as present records indicate, the remnant of the old ice sheet still lingering in Greenland may melt. In melting it will add its locked-up waters to the ocean, raising the sea hundreds of feet above its present level—and thereby flooding the lands along the coast.

II THE OCEAN BEACH

Tides

The ocean runs on moon time. The sun and even distant stars exert a gravitational pull on the earth, but our close neighbor the moon is largely responsible for the ebb and flow of the seas that we call tides. The moon, orbiting the earth, literally lifts the oceans' waters and lets them fall again. As the earth adds its own spinning motion to the moon's pull, two great tidewaves race around the globe during each lunar day. (A lunar day is 24 hours and 50 minutes long. Therefore tides along the Atlantic coast arrive about 50 minutes later each day.)

Although it is the moon that starts the earth's waters moving, the tides that reach our shores are shaped by different coastlines, currents and ocean bottoms. Two

"The Wave"
by Hokusai

high and two low tides advance and retreat on the Outer Lands every day, but from beach to beach the height of the tides and the time of their arrival varies. At Provincetown, tide tables must be read attentively, for the drop between high and low water can be as much as eleven feet. Forty miles away at Nauset Harbor, the tidal range is six and a half feet, while on Nantucket there is little more than a foot difference between the highs and lows.

In addition to these local differences, there are regular changes in the height of the tides. Twice during the lunar month—after the moon is new and after it is full—high tides are higher and lows lower than at other times. Then waves roll up the beach to break against the dunes or, pulling back, uncover broad expanses of sand flats. These spring tides occur when sun, moon and earth are in line and their gravitational forces are pulling together. The "spring" of spring tides has nothing to do with the season. As used here, the word comes from the Old English *springen*, meaning "to leap."

Spring tides last four or five days. A week or so later, when the moon is in its first or last quarter, the tide-waves slacken. Water doesn't travel as high up the beach or pull back as far. There is less difference between each day's highs and lows. This is the time of the neap tides—"neap" going back to the Old English *nep*, which means "unable to advance."

The perigee tides come twice a year, usually in the spring and fall. The highest tides of all, they occur when the orbiting moon is closest to the earth (at perigee) and is either new or full.

Day after day, week after week, the tides leave their mark on the beach. Close to the base of the dunes or cliffs a long line of debris—dry seaweed, driftwood, bleaching shells—shows the extent of the last spring tides. Further down the beach is the line of sea wrack washed up by the neaps. At the water's edge where fresh seaweed glistens on the damp sand and a fish tries desperately to flop back into the ocean, today's tides are spreading out their catch. Distinct as these lines of sea wrack are, they are never exactly identical. For, although they are formed by the tides, they are reshaped by the waves.

Waves

Actually, tides and waves cannot be discussed separately. Tides are waves—the longest waves that oceanographers know of. But the white-crested breakers that swimmers ride in summer are largely caused by winds. Far out in the Atlantic, the wind ruffles the surface of the water. Ripples form, growing higher and longer as the wind increases.

The wind-driven drops of water travel in a circle—not forward as it seems to an observer on the beach, but down and around. And each circle starts another circle moving until great stretches of the ocean are covered by billowing waves.

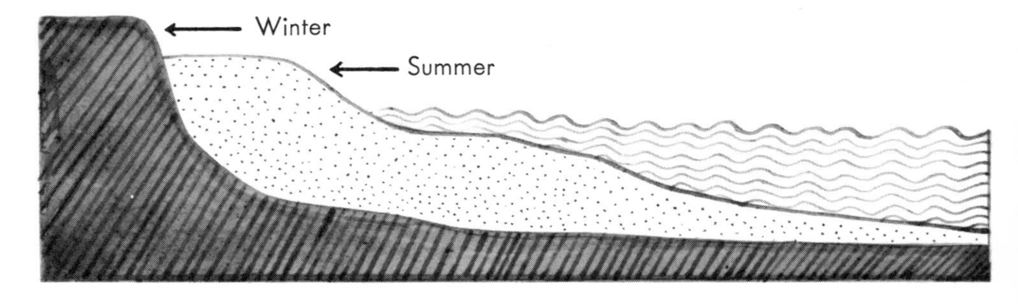

Winter and Summer Beach

As these swells draw close to shore they touch bottom for the first time. Slowed down by the drag of the bottom, a wave's form changes. No longer able to complete its circular motion, it rears up, becoming taller, steeper. Suddenly its white-capped crest topples over. Crashing in the shallow water, it breaks up into a mass of spray and foam.

With a last spurt of energy and a push from the waves behind it, a film of frothy water rushes up the beach. Some of the water sinks into the sand. The rest slides back to sea.

Standing at the water's edge, one can feel the tug of the current caused by the return of the spent wave to the ocean. This is the much-feared undertow. Despite popular beliefs, however, undertow is not guilty of the evil deeds it has been accused of. The seaward trip of the spent wave is a short one. As soon as it reaches deeper water it resumes its circular motion, taking up its old business of being a wave.

The strength of the undertow can be tested by tossing a piece of wood into the water. The wood will bob along, floating out to sea until it reaches the line of breakers. Then, depending on longshore currents, it will probably move parallel to the beach. Waves may wash it to shore a mile away, but they won't take it off to Spain.

This is not to say that the ocean isn't dangerous, particularly for an inexperienced swimmer. Breakers can knock you down. Waves can up-end you or force you into all sorts of unpleasant gymnastics. But as you do cartwheels along the ocean bottom with your mouth full of salt water, remember—it's not the undertow.

Sand

Many centuries ago a Greek philosopher who was impressed by the changing nature of the world around him declared that a man could never step in the same stream twice. He might just as truly have been talking of ocean beaches. For today, tomorrow, next week—the beach is never the same.

Each breaking wave picks up sand from the ocean floor and carries it to shore. And each retreating wavelet transports sand back to sea again. As thousands of waves move billions of sand grains each day, the beach shifts constantly. In a week a new sandbar may appear. In two weeks a boulder at the water's edge may be uncovered or the ribs of an old dory buried under new deposits of sand.

These changes seem accidental, but they follow a definite pattern from season to season. Scientists speak of a winter beach and a summer one. The storm-driven waves of winter cut away the beach, making it steep and narrow. In summer a gentler surf returns sand to

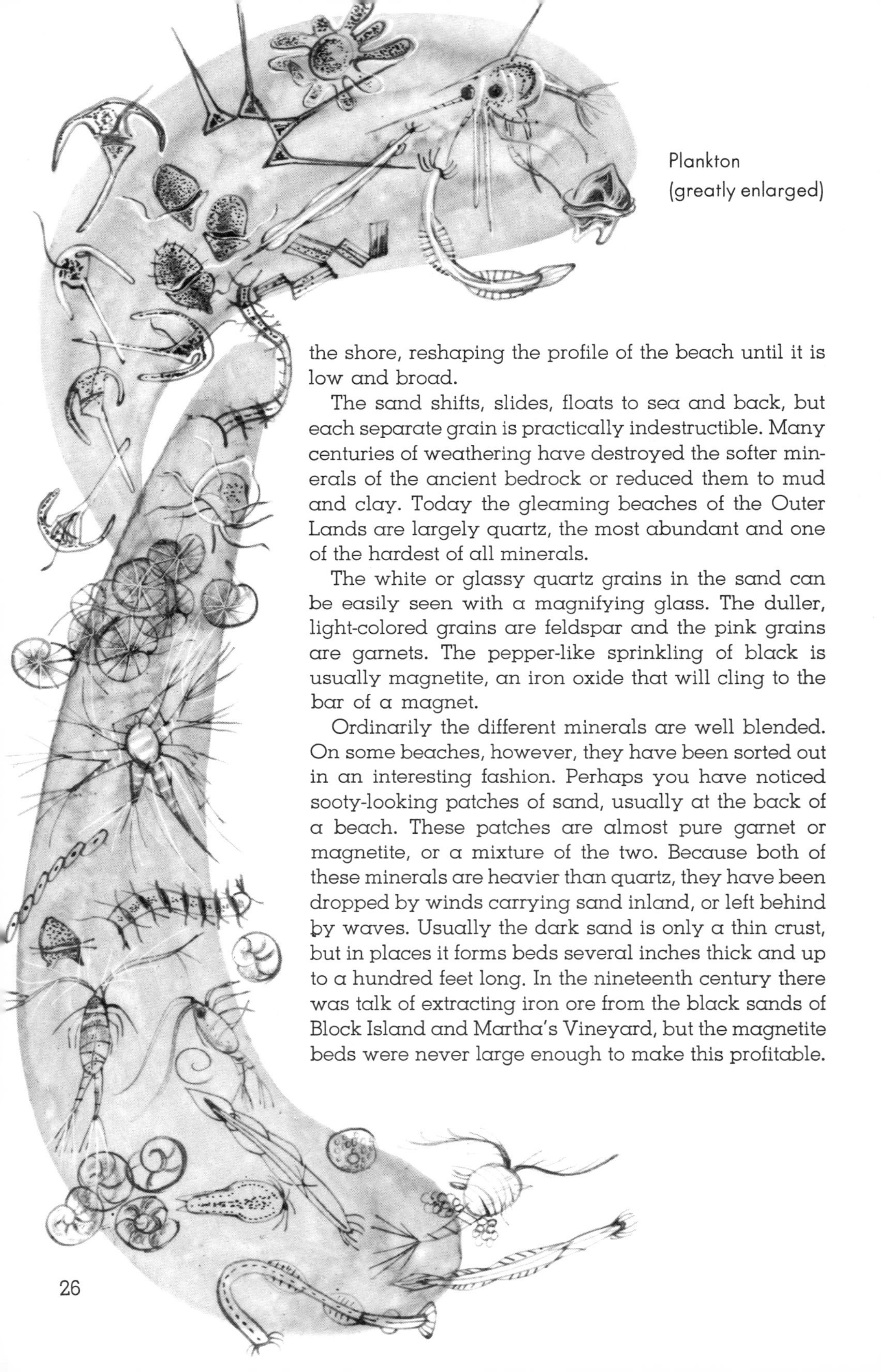

Plankton
(greatly enlarged)

the shore, reshaping the profile of the beach until it is low and broad.

The sand shifts, slides, floats to sea and back, but each separate grain is practically indestructible. Many centuries of weathering have destroyed the softer minerals of the ancient bedrock or reduced them to mud and clay. Today the gleaming beaches of the Outer Lands are largely quartz, the most abundant and one of the hardest of all minerals.

The white or glassy quartz grains in the sand can be easily seen with a magnifying glass. The duller, light-colored grains are feldspar and the pink grains are garnets. The pepper-like sprinkling of black is usually magnetite, an iron oxide that will cling to the bar of a magnet.

Ordinarily the different minerals are well blended. On some beaches, however, they have been sorted out in an interesting fashion. Perhaps you have noticed sooty-looking patches of sand, usually at the back of a beach. These patches are almost pure garnet or magnetite, or a mixture of the two. Because both of these minerals are heavier than quartz, they have been dropped by winds carrying sand inland, or left behind by waves. Usually the dark sand is only a thin crust, but in places it forms beds several inches thick and up to a hundred feet long. In the nineteenth century there was talk of extracting iron ore from the black sands of Block Island and Martha's Vineyard, but the magnetite beds were never large enough to make this profitable.

The Wanderers

Englishmen visiting the Outer Lands early in the seventeenth century marveled at "the abundance of Sea-Fish...Sculles of mackerall, herrings, Cod and other fish that we dayly saw as we went and came from the shore were woonderfull." And after the lands were settled, the cry of "Whale off!" was enough to bring every able-bodied man to the shore.

Today's warmer ocean is driving the cod and sea herring further and further north, but from Jones Beach to Provincetown mackerel and bluefish, swordfish, tuna and striped bass are still abundant, as the fishing boats on the horizon and the fishermen on the beaches will testify. Harbor seals raise their pups on offshore islands and an occasional whale pays a visit to the beaches—only to be towed back to sea by the Coast Guard.

Drifting through the blue-green waters are billions of minute plants and animals—one-celled diatoms and radiolaria, microscopic copepods, transparent arrow-worms. These plankton (from a Greek word meaning "wandering") are the basic food supply of the sea and the narrow strip of land that borders it. Just as

> Great fleas have little fleas
> Upon their backs to bite 'em
> And little fleas have lesser fleas
> And so *ad infinitum*

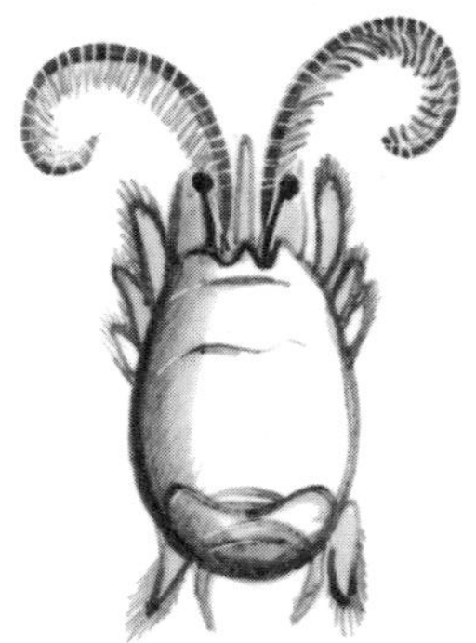

so copepods feed on diatoms and herring feed on copepods and mackerel feed on herring and tuna feed on mackerel—until humans, those sea-going land animals picnic on tuna-fish sandwiches.

Mole Crabs and Sand Hoppers

The breaking waves and endless stretches of sunbaked sand—the very elements that delight vacationers—make the ocean beach the barrenest region of the Outer Lands. The surf scours the surface of rocks offshore, allowing only a fringe of seaweed and the hardiest snails and barnacles to survive. Along the beach no more than a handful of animals and a few stunted plants can find food and anchorage in the shifting sands.

Mole Crabs

In the churned-up waters at the edge of the beach, you may catch a glimpse of black eyes and whiskered faces which appear and fade away so quickly that you're not sure if you have dreamed them. These are Mole Crabs, true crabs closely related to Hermit Crabs. Egg-shaped, sand-colored, they bury themselves in the sand with only eyes and antennae showing. Facing the ocean, they wait until a spent wave returns them to sea. Then, using their feathery antennae as nets, they "fish" in the inch-deep water. During the brief period of calm before the next wave rushes in, they transfer their plankton catch to their mouths.

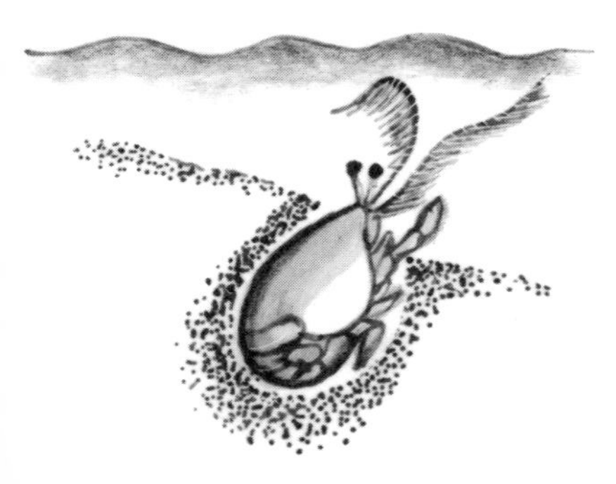

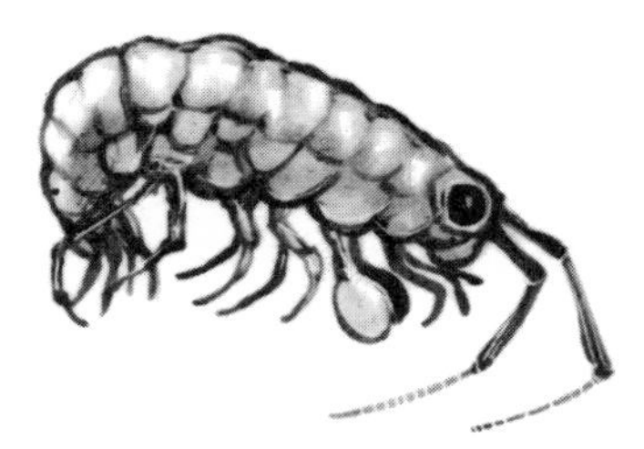

Sand Hoppers

If you can scoop up a few of these crabs—it's not easy—carry them to dry sand for a better look at them. One will play dead while another, traveling backward, scurries off toward the sea. On wet sand they vanish, digging under with tails and hind legs faster than you can say "Mole Crab."

Living at the edge of the surf, Mole Crabs follow the tides. Whole colonies travel together, leaving their burrows at the same time to ride up the beach on a wave. They dig in side by side, until an ebbing tide forces them to pull up stakes and move once more.

Mole Crabs mate in the spring. By mid-summer most of the females have masses of orange eggs under their tails. The newly hatched young are often carried away by longshore currents to start new colonies far from their parent beach.

While Mole Crabs "fish" in the sea, other creatures feed at the tide line. Each wave leaves a ripple of sand and a sampling of plankton on the beach. As the water recedes, clouds of Sand Hoppers appear. So small that they are scarcely visible, they bounce up and down, hunting for plankton and other refuse that is washed ashore. Hundreds congregate on a dead fish or under piles of decaying seaweed.

Despite their nickname, "beach flea," Sand Hoppers are not insects but amphipods, marine animals related to shrimps and crabs. Several different kinds of Sand Hoppers live on the beaches and in shallow waters close to shore. Appearing pearl-white in color, they are

Sanderlings

actually olive-green, pink, or spotted with red when examined with a magnifying glass. Seldom more than half an inch long, they have flattened bodies and enormous eyes. They leap across the sand with the help of their tails and rear legs. In the water they swim on their sides.

Some Sand Hoppers dig burrows high up on the beach where they hide during the day, coming out at night to feed. Although they hunt along the tide line, these hoppers of the upper beach drown if they stay in the water too long. Before sunrise each day they dig fresh burrows, peppering the sand with their holes.

Holes on the upper beach are usually empty Sand Hopper burrows. Closer to the water, holes of similar appearance are formed by uprushing waves. As the water sinks into the sand it displaces the air between the sand grains. Forced upward, the air forms little craters in the surface of the beach.

Birds

Terns fish in the offshore waters. Piping Plovers nest above the tide line, hollowing out the loose sand and lining the depression with pebbles and shell fragments. Countless thousands of land and shore birds stop over on the beaches during spring and fall migrations. However, the characteristic birds of the ocean beach are the Sanderlings.

These small sandpipers fly to the shores of the Arctic Ocean for nesting, but they spend a large part of the year on the Atlantic coast. Called "surf snipes," they follow the retreating waves. With their sharp bills they pick daintily at the minute creatures of the backwash, or probe beneath the sand for Mole Crabs—which are as big as the birds' own heads. Seldom alone as they trot across the sand, they walk and run, wheel and fly in unison, like the members of a well-trained ballet corps. Hunting from dawn until sunset, the Sanderlings take catnaps on the beach during the day. Some squat on the damp sand while others balance on one leg, swinging around like weather vanes as the wind blows.

Plover chicks

Bank Swallows

Although they are not usually thought of as shore birds, colonies of Bank Swallows nest in the bluffs behind the beach, digging long tunnels in the compacted gravel. When Henry David Thoreau walked along Cape Cod's Great Beach more than a century ago he counted two hundred Bank Swallow holes in the cliffs below the lighthouse at Truro. Even though the cliffs have been eaten away by the surf, hundreds of Bank Swallows still return to Truro each year. Darting in and out of their tunnels, they refurnish them with grass and feathers and drop down to the beach to feed.

Beachcombing

Furry tent caterpillars are blown onto the beach in the spring. Tiger beetles hunt along the lines of sea wrack in summer. Migrating Monarch butterflies follow

Torpedo Ray

the shores in the fall. Even raccoons and foxes and white-footed mice pad across the sands after dark.

Most visitors, however, come by water. As the tide ebbs, fish are sometimes left high and dry. The commonest victims of the tides are the Skates. Strange-looking flat fish with eyes on their backs and broad winglike fins, they live in shallow water. With mouths conveniently situated on their undersides and jaws built for grinding, they dig up clams and break open the hard shells to get at the meat inside.

Unlike most fish who lay thousands of tiny eggs, Skates lay large yolk-filled ones, as big as pigeons' eggs. Each egg is enclosed in a black parchment-like pouch which the parent Skate anchors to seaweed or a submerged rock. When the young Skate hatches after almost a year of incubation, its empty pouch, often called a "mermaid's purse," is washed ashore.

Skates are harmless, but if you find their relative, the Torpedo Ray, on the beach, be prepared for a shock. These believe-it-or-not fish generate as much as 220 volts of electricity in their flat bodies. They can deliver charges with sufficient amperage to stun a man, but not enough to kill him. (Dead rays, however, are safe to handle.) Although rays are warm-water fish, they are often stranded on the beaches of Long Island.

Aside from the Skates and their black egg cases, an hour's beachcombing along the ocean is not likely to yield enough to fill a bathing cap. You will see battered clam shells and bright pebbles whose colors fade as they dry, silvery driftwood and smooth stones that

Skate

beg to be skipped across the water—and beer cans and plastic jugs that no amount of sun and surf can make lovely. Occasionally a storm will bring in unfamiliar spiral shells from deeper waters. Or seaweed ripped from its moorings will carry with it small jellyfish and colonies of minute animals. But for perfect shells and the animals who live in them, you will have to visit the beaches of the bays and sounds.

Skate egg case

Ladder Shell

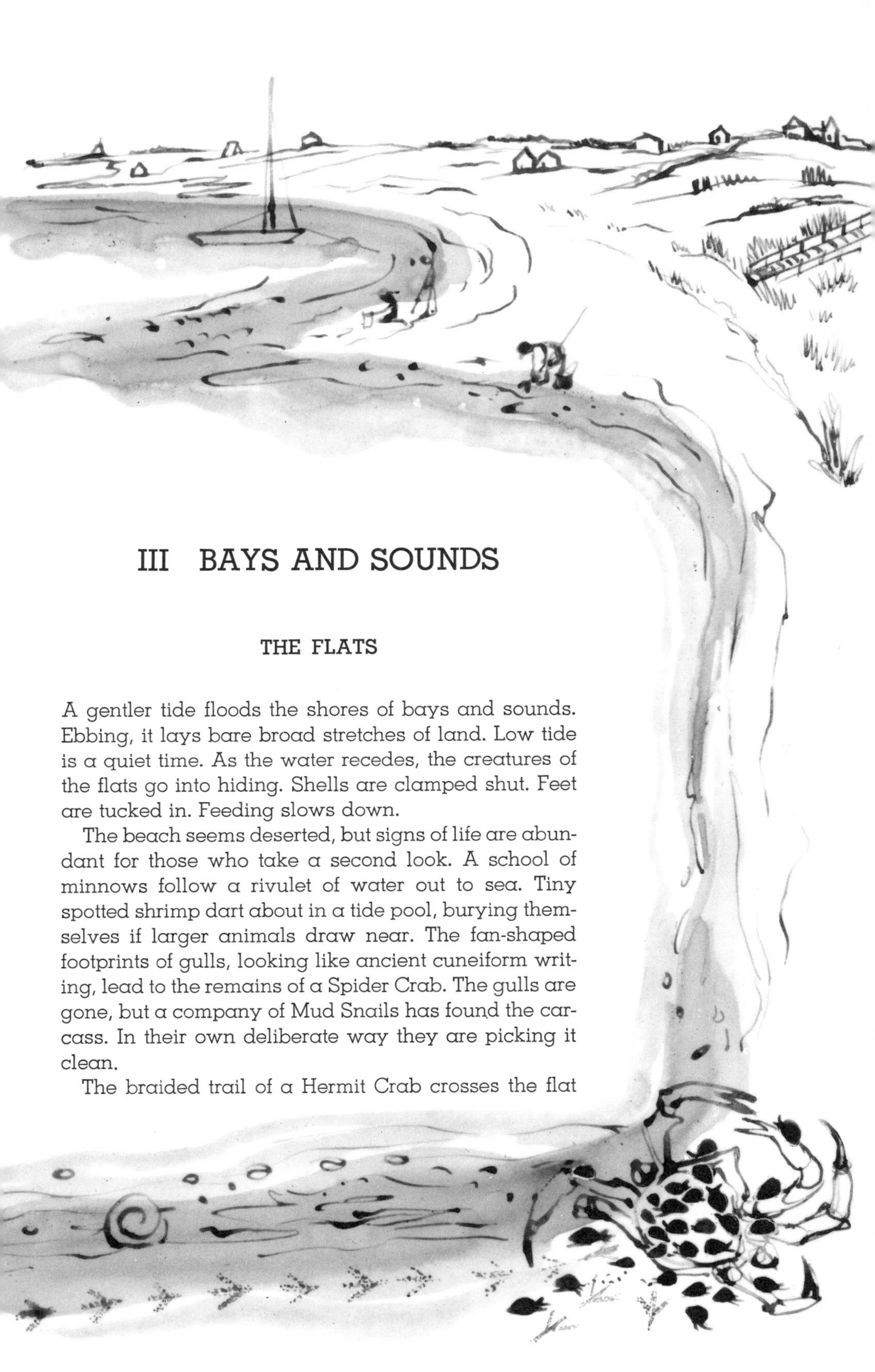

III BAYS AND SOUNDS

THE FLATS

A gentler tide floods the shores of bays and sounds. Ebbing, it lays bare broad stretches of land. Low tide is a quiet time. As the water recedes, the creatures of the flats go into hiding. Shells are clamped shut. Feet are tucked in. Feeding slows down.

The beach seems deserted, but signs of life are abundant for those who take a second look. A school of minnows follow a rivulet of water out to sea. Tiny spotted shrimp dart about in a tide pool, burying themselves if larger animals draw near. The fan-shaped footprints of gulls, looking like ancient cuneiform writing, lead to the remains of a Spider Crab. The gulls are gone, but a company of Mud Snails has found the carcass. In their own deliberate way they are picking it clean.

The braided trail of a Hermit Crab crosses the flat

ribbon-like paths of the Mud Snails. Nearby a moving hump of sand marks the progress of a Moon Snail, hunting below the surface for a meal. Closer to the water plowed furrows reveal the hiding place of a Horseshoe Crab.

As your eyes adjust to the small scale of life on the flats, you notice tubes, shells and conical mounds protruding from the sand. A jet of water wets your ankles, calling attention to holes, round and wedge-shaped, large and small. Beneath the surface on which you stand there is a maze of tunnels, burrows, passageways, where animals live in darkness, waiting for the tide to turn.

Clams

As the psychologists would say, it is hard to relate to a clam. Headless and brainless, clams live in ways vastly different from our own. When scientists began to classify the world around them, they divided the animal kingdom into two broad groups. With understandable bias, they put animals like themselves—mammals, birds, fish, frogs—into one group. These vertebrates make up about 5 per cent of the animal kingdom. The remaining 95 per cent are lumped together as invertebrates—the animals without backbones.

Clams, along with all the other creatures of the shore except birds and fish, are invertebrates. Further classified according to their body plan, clams are mollusks,

Soft-shell Clam

bivalves, and pelecypods. Which means that they are soft-bodied animals with two shells and a foot that, with a stretch of the imagination, can be described as hatchet-shaped.

The familiar squirting clam of the flats is the Soft-shell Clam, known also as a Steamer or Long Neck Clam. The clam's "neck" is actually a muscular double-barreled siphon. One tube of the siphon takes in water containing the plankton on which the animal feeds. The other returns water and wastes to the sea.

The clam lives under the sand in a vertical position, with its "neck" pointing upward and its foot down. Its soft body—gills and heart, digestive and reproductive organs—is wrapped in a thin membrane called a mantle. The mantle is able to take lime from the sea for shell-building. Since a clam grows throughout its life, slowing down in winter, it has curved growth lines on the outer surface of its shell. Like the annual rings of a tree, they give an approximate idea of the clam's age.

During the summer adult clams discharge millions of eggs and sperm into the water. Fertilization takes place when a minute sperm accidentally comes in contact with an egg. In a few hours the fertilized egg develops into a microscopic swimming creature that lives near the surface of the water. After less than a fortnight, the clam's swimming days come to an end and it drops to the bottom. There it anchors itself to sand grains or seaweed by means of a thin thread called a byssus. From time to time the tiny clam casts off this thread and creeps slowly across the flats. As it grows larger it stops moving about and begins to dig down. By the time the clam is two years old—and about two inches long—it loses its ability to crawl. Digging deeper under the sand, it stays put for the rest of its life—which may be as long as twelve years.

As the tide rises, clams of all ages reach up their siphons to take in streams of food-filled water. At low tide they pull back their siphons, leaving holes on the surface of the sand. Footsteps or the vibrations caused by a clam rake will make a clam close its shell and retract its siphon further, squirting water as it does so. A Soft-shell Clam does not wriggle away from you when you dig. It is only the pulling back of its siphon that makes this seem to happen.

The Hard-shell Clam of Long Island, the Quahog of New England, the Little Necks and Cherrystones listed on restaurant menus are all one and the same animal. The Quahog has a heavier shell and a shorter "neck" than the Soft-shell Clam. Because of the stubbiness of its siphon, it lives just below the surface of the sand, at the edge of the flats or in deeper water offshore.

Developing in much the same way as Soft-shell Clams, Quahogs can live twenty to twenty-five years. As tender two- and three-year-olds, they are Little Necks and Cherrystones, fated to be eaten on the half-shell. Growing older, they are likely to end up in chowder.

In addition to their place on the dinner table, Quahogs have a place in history. For hundreds of years, Indian tribes traveled to the shore each summer to

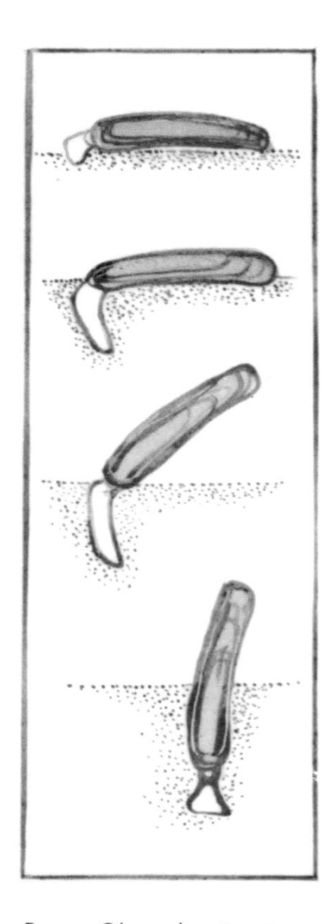

Razor Clam digging in

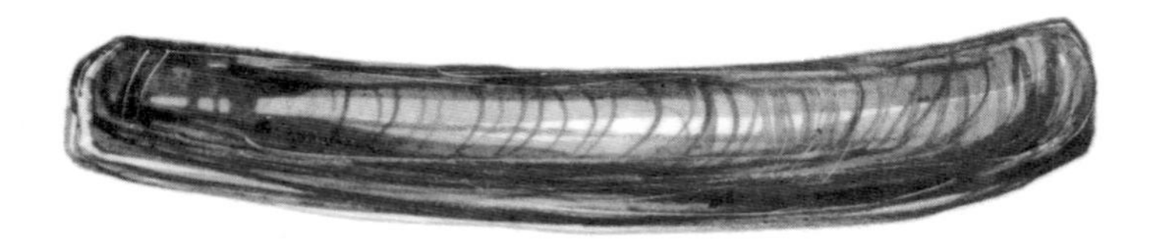

Razor Clam

collect shells for wampum beads. To the Indians of New York State, Long Island was known as Si-wan-aki, the Land of the Shells. Originally their beaded necklaces and belts were only for ceremonial use. After white traders introduced the idea of using wampum as money, the purple beads made from the edge of Quahog shells had twice the value of white shell beads.

The Razor Clam, shaped like an old-fashioned straight razor and with a sharp cutting edge to its shell, is another inhabitant of the flats. With a short siphon and a long foot, it lives close to the surface. At low tide, its shell sometimes protrudes above ground. Move quickly if you want to catch one then, for a Razor Clam is fast. Its pointed muscular foot reaches deep into the sand, pulling down the thin, streamlined shell. Once underground, the tip of the foot swells, forming a rounded anchor which grips the sand. If you grab from above while the clam pulls from below, you will probably lose the round.

Should you win, however, put the Razor down on the flats and watch it operate. Its mobile foot pokes out of the end of the shell and curves into the sand. Suddenly, clam and shell leap upright—and just as suddenly the animal vanishes underground, leaving only an oblong hole to show where it has gone.

If you speak of a person "shutting up like a clam" you are not talking about a Razor. Unlike Quahogs whose shells seal so tightly that they can live for days out of water, a Razor can't "clam up." Its shell is open on both top and bottom, leaving it an easy mark for

Surf Clam

gulls and other predators. A Soft-shell Clam can't clam up either, although it does better than a Razor. In England where it has been introduced from our shores it's called "the gaper."

The Surf Clam, or Sea Clam, is the largest on Atlantic beaches. Buried less than an inch under the sand, it lives beyond the low-water mark and in deeper water. Clam-diggers find it at the outer edge of the flats during low spring tides. The live animal's shell is brown, bleaching to a chalky white when it washes up on the beach.

Familiar now as the ashtray shell of summer cottages, Surf Clams were prized by Colonial housewives. They served the meaty animals in clam pies and used the big oval shells in place of soup ladles. Known as "skim-alls," the shells were standard kitchen equipment all over the Outer Lands.

On beaches where a thin layer of sand covers an ancient clay or peat bed you can probably find False Angel Wings. Unmistakable clues to their presence are pairs of holes raised slightly above the surface of the beach. These paired holes are left by the animals' long slender siphons, which are divided instead of joined together. The clams live several inches underground, in burrows that they have dug in the stiff clay.

False Angel Wings and the Fallen Angels which sometimes share their clay beds are known as boring clams, a name that would seem less humorless if you could see them in action. Their fragile-looking shells are decorated with raised toothlike ridges. By rotating

Pandora
Chestnut
Gem
Ark
Tellin
Quahog
Fallen Angel
False Angel Wing
Angel Wing

CLAMS

the shells and moving slowly up and down, the boring clams are able to chip out deep grooves in wood, concrete and even rock. The true Angel Wings are also boring clams. Although occasionally found on the beaches of the Outer Lands they are more common further south.

The catalog of clams on the flats and in shallow water close to shore is a long one. Scattered along the beaches are Ark Shells, often called Bloody Clams because, unlike most mollusks, they have red blood. Equally common are the red-brown Chestnut Clams, the tiny Gem Clams and the delicate and beautiful Pandoras and Tellins. In spite of their wide range of sizes and shapes, all of these creatures are bottom-dwellers who filter food from the water through their gills.

False Angel Wings

Snails

Snails are mollusks whose soft bodies are covered by one-piece spiral shells. Because they seem to travel on their stomachs, they were long classified as gastropods, from the Greek, "stomach-footed." Actually a snail's stomach, as well as its other internal organs, is wrapped in a mantle similar to a clam's. They fit neatly inside the coiled shell, while head and foot meet at its opening.

The architecture of a snail's shell is not only a thing

Mud Snails (actual size)

of beauty but a precise bit of engineering. Light enough to be carried around, the shell is both fortress and shelter. If disturbed, the snail pulls in its head and foot. A flat horny shield on the hind end of the foot acts as a trap door. This operculum, as it is called, plugs up the shell opening, protecting the snail's soft body and preventing it from drying out between tides. As the shell grows, the operculum grows too, so that it always forms a tight seal. Circular or oval, depending on the shape of the snail shell, an operculum has wavy growth lines similar to those on a clam shell. Although most marine snails have opercula, fresh-water and land snails usually do not.

Most of the snails of the flats are meat-eaters. When the sensitive tentacles of Mud Snails pick up the odor or taste of dead fish, they glide across the sand on their flat feet. Coming from all directions, at a snail's pace (which has been clocked at two inches per minute), they swarm over their prey. When feeding, the Mud Snail stretches out its proboscis, a flexible snout which waves from side to side like an elephant's trunk. At the tip of the proboscis is a remarkable ribbon-like tongue. Called a radula, it is equipped with rows of sharp teeth. By pulling it back and forth, the animal scrapes and shreds its food. Although they are normally scavengers, Mud Snails can also use their radulas to drill through the shells of clams and even their snail neighbors.

The smooth globe-shaped shell and enormous foot of the Moon Snail permit it to burrow under the sand—

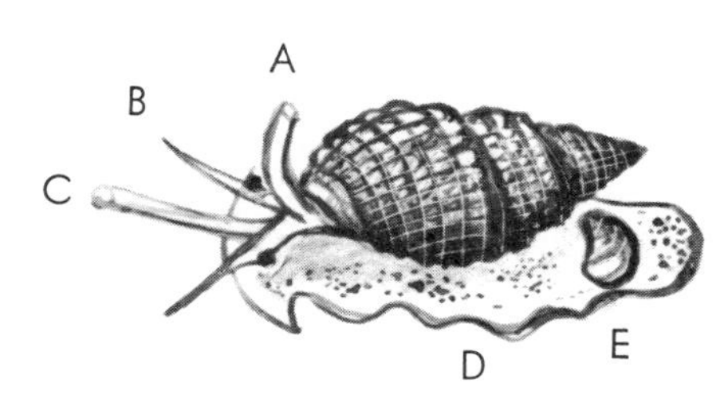

A. Siphon
B. Tentacles
C. Proboscis
D. Foot
E. Operculum

as deep as twelve inches—in pursuit of clams and other mollusks. The oversized foot has a special flap in front for digging, and a rear flap which covers the animal's head. Although it doesn't seem possible, the foot, flaps and all, can be packed into the shell. In order to do so, the snail squeezes out quantities of water, until its foot shrinks to a stowable size. It cannot remain closed in its shell for long, because its body is wedged in so tightly that it can scarcely breathe.

Plowing under the sand, the Moon Snail wraps its foot around a helpless clam. Then it goes to work with its radula. After drilling a round, countersunk hole in the clam shell, it reaches in with its proboscis and slowly eats its victim. In a laboratory aquarium, a Moon Snail spent more than two days drilling and devouring a smaller snail.

The whelks are the most spectacular of the hunting snails. The Knobbed Whelk, sometimes nine inches

long, is the largest mollusk of the Outer Lands, with the Channeled Whelk running it a close second. Traveling across the flats with its broad foot just below the surface, a giant whelk has little difficulty in capturing its prey. Although it can drill holes with its radula, it also uses its heavy shell as a hammer and, with a series of blows, cracks open the shell of its victim.

The smaller Waved Whelk which is sometimes found on shore is actually a creature of deeper waters. Considered a pest by lobstermen because it steals bait from their traps, it is a popular dinner dish in Europe.

All of these snails construct elaborate egg cases for their young. Working under the sand during the spring and summer, a female Moon Snail prepares for egg-laying by giving off a film of mucus. As sand grains become trapped in the sticky film, she starts to lay eggs. The eggs, emerging in a continuous ribbon, are caught in the mucus too. A growing collar of sand, with eggs glued to its underside, forms around the animal, taking its shape from her foot and shell.

When all her eggs are laid, the snail crawls away, leaving the egg case on the flats. Leathery when it is wet, the sand collar becomes brittle if it dries. After the eggs hatch, the crumbling case is often washed up by the tides. Sometimes you can still see eggs fastened to its underside.

Knobbed and Channeled Whelks enclose their eggs in tough parchment capsules which are fastened together by a string of the same material. The eggs are laid under the sand and the twisted string of capsules, which may be a yard long, is pushed up to the surface as it is completed. Each capsule contains about two dozen eggs. The whelks hatch inside the parchment cases. When they escape through a thin spot on the capsule's outer edge, they are tiny copies of their parents. Occasionally an egg-string on the beach will contain some of these minute snails.

The rounded egg cases of the Waved Whelk are joined together to form an irregular ball. Although there are hundreds of eggs in each pea-sized case only a few snails survive. Carnivorous from the day they hatch, the more vigorous babies eat their weaker brothers. Sailors of an earlier day discovered that if these wet egg cases were rubbed between the hands they would lather as soap does. Old-timers along the shores still speak of them as "sea-wash balls."

The Boat or Slipper Snail breaks all the rules of snaildom. Its shell is boat-shaped instead of spiral, with a broad deck inside. In addition, the animal stays put, as clams do, instead of moving around. After swimming for two or three weeks, a young Boat Snail looks for a place to settle down. Using its foot as a sucker, the snail fastens to a pebble or an empty clam shell, or chooses a shell occupied by a Hermit Crab and gets a free ride. It is able to move until it is half-grown, but after that it settles down permanently, cementing itself in place. As succeeding generations of Boat Snails search for home sites, they often cling to their fellows. Sometimes you can find a dozen snails piled one on top of the other, with only the bottom snail anchored to a stone.

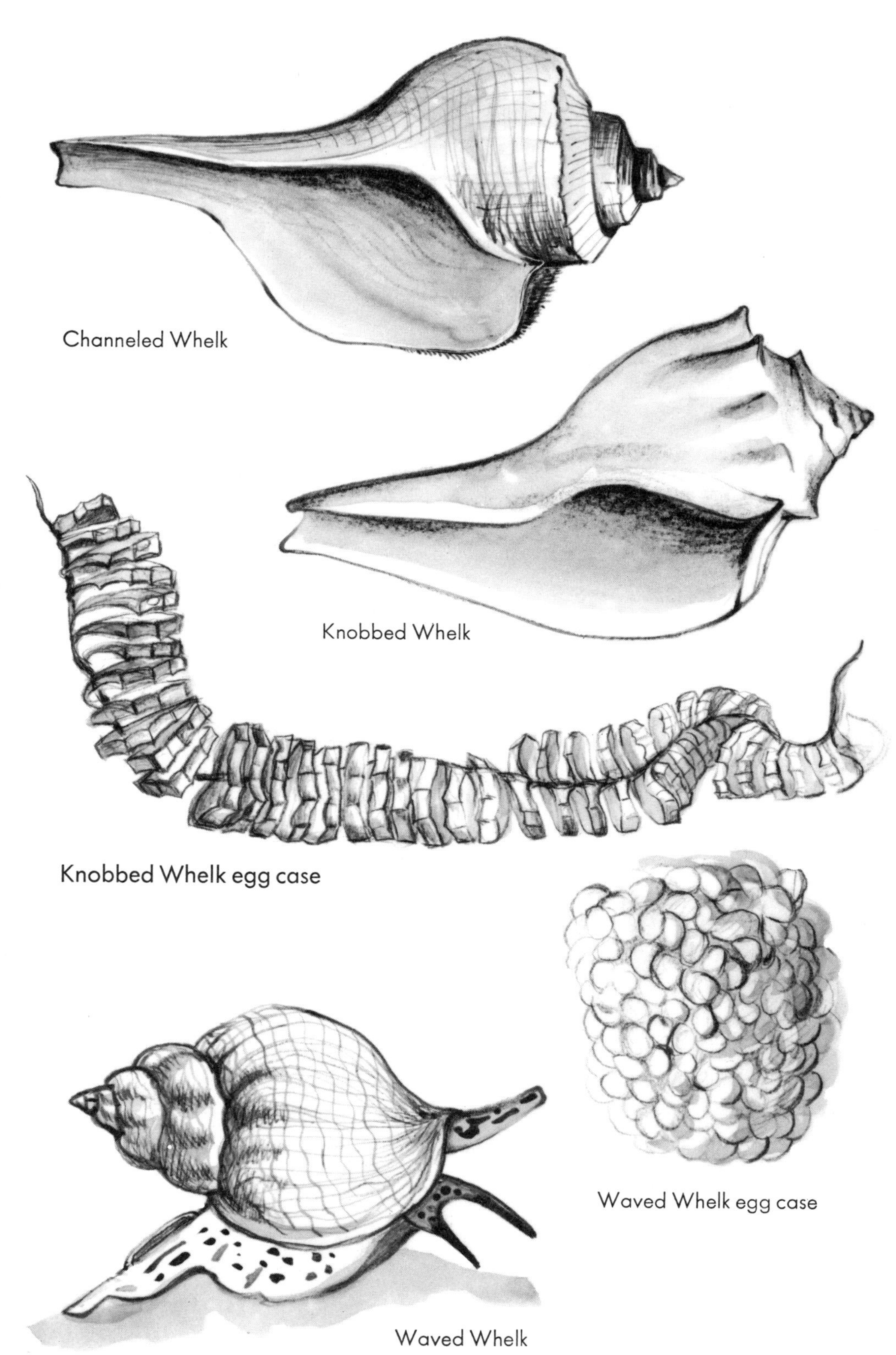
Channeled Whelk

Knobbed Whelk

Knobbed Whelk egg case

Waved Whelk egg case

Waved Whelk

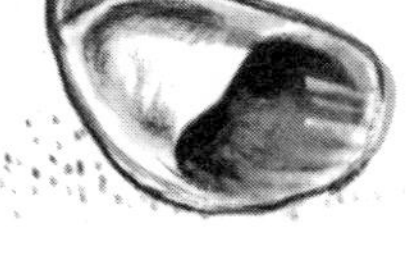

Boat Snails

When the tide rises, the Boat Snail lifts its shell a fraction of an inch to permit water to reach its gills. At low tide, with its shell clamped down tightly, it continues to feed on the bits of plankton it has collected.

Boat Snails change their sex as they grow. They start out as males, then become females, and sometimes males again. These changes depend in part on the sex of their neighbors. The snails at the bottom of a long chain are always females. Those in the middle are usually in the process of changing sex, while the younger, smaller snails on top are males. Although it seems strange, this ability to change sex is fairly common among invertebrates, particularly among sedentary animals who cannot travel around to find mates.

Hermit Crabs

Seen head-on, a Hermit Crab (Plate 1) is a typical crab, with a hard outer skeleton and menacing claws. From the rear, it is unarmored and defenseless. To protect its soft abdomen, the crab uses an empty snail shell as a portable home.

During a long evolution on the shore, the bodies of Hermit Crabs have become adapted to their borrowed homes. Their curving abdomens fit neatly into the spiral shells. Hind legs, no longer used for walking, have become hooks which clasp the shell's central column, making it almost impossible to pull the animal out. (One scientist found that if he drilled a small hole in the rear of a Hermit's shell and poked at the crab

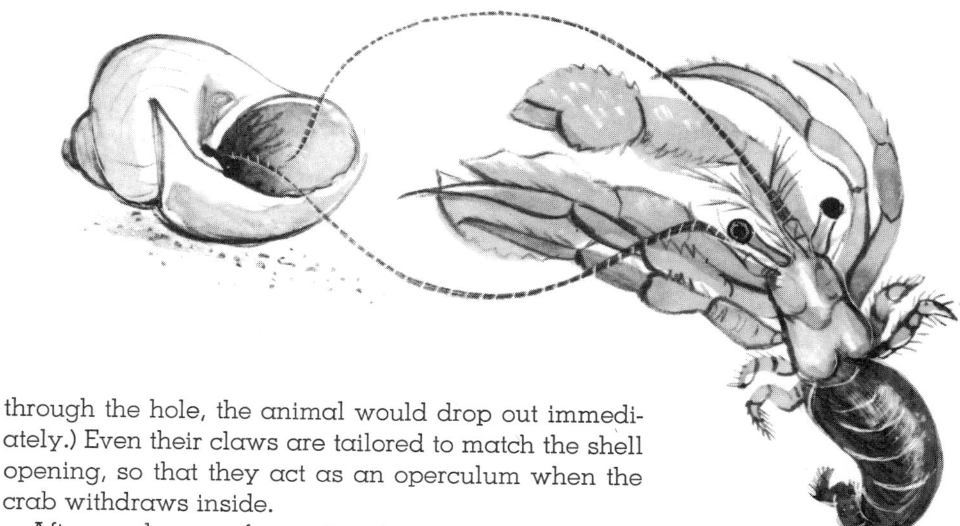

through the hole, the animal would drop out immediately.) Even their claws are tailored to match the shell opening, so that they act as an operculum when the crab withdraws inside.

After weeks as a free-swimming creature, a young Hermit Crab takes up residence in a shell and scuttles around on the floor of the sea to find food. When eating, it uses its claws almost as if they were knife and fork. As it holds on to the remains of a clam with its large right claw, its left claw cuts the meat and carries it to its mouth.

Hermit Crab
without shell

Growth confronts the Hermit Crab with a lifelong housing problem. Unable to enlarge its shell, it constantly searches for a new home. Any shell it sees must be investigated carefully. The crab rolls the shell over, feeling it with its claws and tapping it with its long antennae. If the result of the inspection is satisfactory, the crab darts out of its old shell and into the new one. After slipping it on for size, it may pop back into its old home again.

In a well-stocked tide pool, a Hermit Crab never seems to tire of this game of musical shells. As pugnacious as its larger crab relatives, it battles over a shell, fighting not only a fellow Hermit for possession, but even trying to oust a live snail.

In late spring, Hermits wander across the flats in pairs. The larger of the two crabs is a male who has grabbed the shell of a female with his claw and is dragging it around. After mating, the female carries her eggs inside her shell, in the curve of her abdomen.

When they are ready to hatch, she crawls part-way out of the shell to brush them off into the water with a hind leg.

Two species of Hermit Crabs are abundant on the shores of the Outer Lands, one living in the shells of Mud Snails and Periwinkles and the other in Moon Snail and whelk shells. The big Hermits are seldom alone because young barnacles and Boat Snails, seeking a place to attach, settle down on their shells. These hitchhikers grow until sometimes the Hermit can barely shuffle along under their weight.

In addition, a Hermit's shell is often covered with a pink or purple fuzz. With a magnifying glass you can see that this fuzz consists of tiny stalked animals of different shapes and sizes. These are hydroids, colonial animals related to jellyfish. Some members of the hydroid colony do the feeding for the whole group, while others are concerned with reproduction or defense.

Further south along the Atlantic, a large sea anemone lives on Hermit shells. Both landlord and tenant benefit from the relationship, the anemone feeding on scraps of meat dropped by the Hermit, and in turn providing the crab with protection and camouflage. When the Hermit is ready to transfer to a new shell, it strokes the anemone's stalk until the flower-like animal moves too. This kind of mutual relationship is known as commensal—literally "eating at the same table."

Hermit Crab
with sea anemone

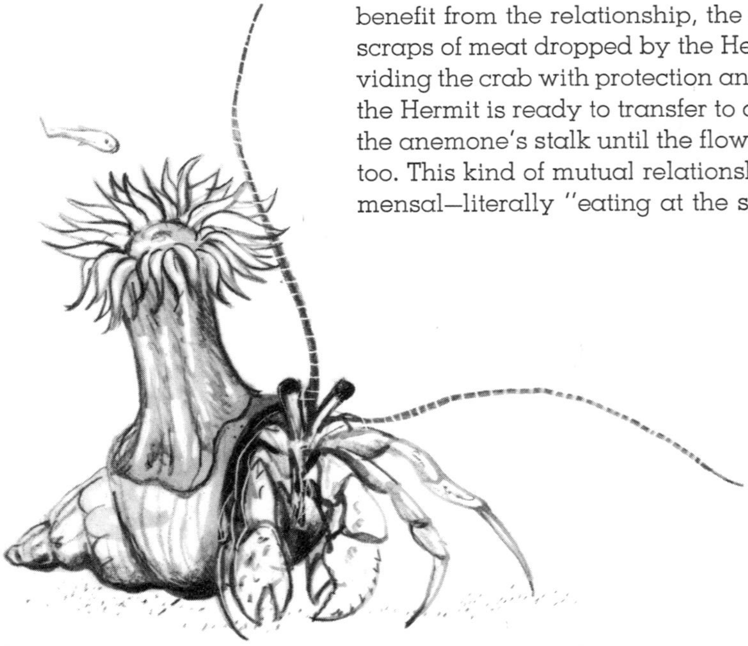

Worms Can Be Beautiful

As any clam-digger knows, not every hole on the flats leads to a clam, nor do two holes mean two clams. More often than not a clam rake will turn up worms. Three groups of worms live along the shore, but the most interesting are the segmented worms, known as annelids, from the Latin, meaning "ringed." (Earthworms and leeches are also annelids.)

Although "worm," often coupled with the word "lowly," conjures up unpleasant images, these marine worms are by no means at the bottom of the evolutionary ladder. They are lowly when compared to the readers of this book, but their bodies and behavior are admirably adapted to the tidal world in which they live. Viewed without prejudice, they can seem surprisingly beautiful.

The Clam or Sand Worm (Plate 2) which fishermen use for bait is brilliantly iridescent, the males blue-green and the females green, tinted with brick red. Each segment of their bodies except heads and tails is equipped with bristly parapods ("side feet") which help them to dig and swim.

Clam Worms spend their days in burrows in the sand. At night they swim through the shallow water with graceful rippling movements. Hunters, they feed on clams, shrimp, and other worms—and are a favorite food of fish. Their powerful scissor-like jaws can in-

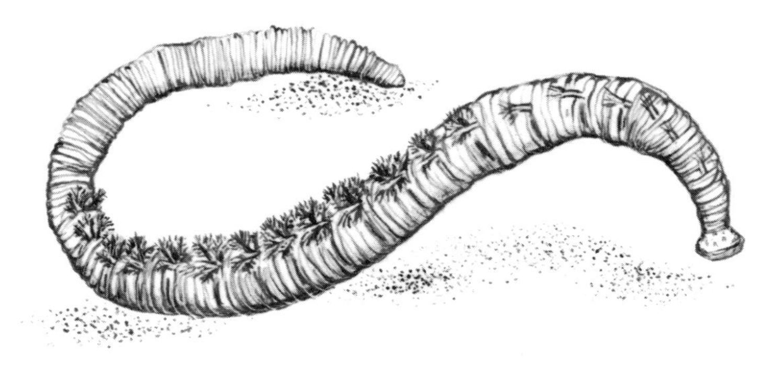

Lugworm

flict a painful bite. If you pick one up be sure to hold it behind its head.

The mating habits of Clam Worms and their relatives are spectacular. At certain times each year—times that are related to phases of the moon—great numbers of them leave their burrows to spawn. The Clam Worm's scientific name, *Nereis*, after the sea nymphs of Greek mythology, seems almost appropriate when thousands (in some places, millions) of these sea creatures meet to perform a nuptial dance. Spiraling rapidly through the water, the males shed their sperm, while females circle around them, releasing eggs. This striking water ballet goes on for hours, leaving the sea filled with fertilized eggs.

Fringed Worms (Plate 2) live in muddy places, under stones and in tubes that they build. While they burrow, their brightly colored tentacles remain on the surface. These threadlike twisting tentacles are actually gills that collect oxygen for the buried worm. Small Blood Worms (Plate 2) also live in the mud. Their tentacles, which are longer than their bodies, are blood-red.

The Lugworm, another bait worm, eats itself into house and home. As it burrows, it swallows most of the sand it dislodges. It digs a U-shaped tunnel with "entrance" and "exit" shafts that end at the surface of the beach. Living at the curved bottom of the U, the worm crawls up one shaft to feed and backs up the other to discharge wastes. Its castings, resembling those left by earthworms, appear on the flats as coiled strings of sand. A saucer-shaped depression a few inches away marks the head end of the burrow. The rising tide washes away the castings and fills in the depressions, bringing fresh sand and water to the buried animal. Only one other creature leaves sand castings on the flats. This is the Acorn Worm which, despite its name and wormlike shape, is not a true worm. Less common than the Lugworm, it is abundant on some beaches.

In their dark underground world, Lugworms carry on a regular cycle of activities. Eating, expelling wastes and then resting, they chug up and down their burrows with clocklike precision. One scientist compares their alternate feeding and resting with a refrigerator, which also has spurts of activity followed by a long rest period. The refrigerator, of course, is controlled by a thermostat which switches on and off as the temperature changes. In their unchanging environment, Lugworms seem to be paced by their own internal clocks.

Several other worms build sturdy tubes under the sand. The Parchment Worm's U-shaped tube is made of a tough paper-like material. The tube itself is almost an inch in diameter, but its twin chimneys which are visible above ground are the size of drinking straws. Once its tube is completed, the worm never leaves home. Using its parapods as paddles, it keeps a steady stream of water flowing in and out.

Besides running water, the Parchment Worm's tube has other homelike features—boarders and a light.

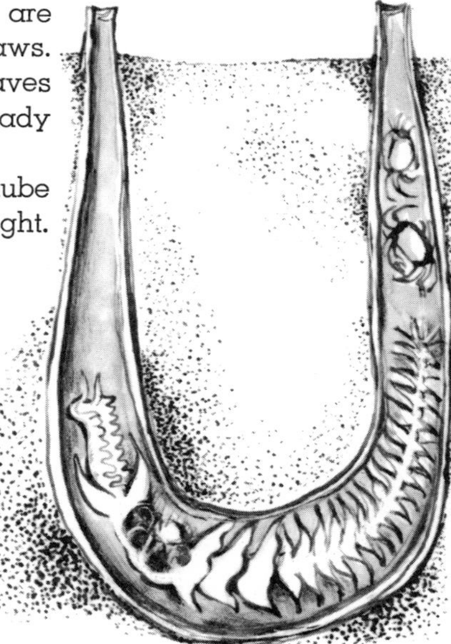

Parchment Worm

Young Pea Crabs often wash in with the incoming current. By the time they are full-size—the size of a pea—they are too big to escape through the tube's narrow chimneys. They live out their lives side by side with the worm. More puzzling than these crabs-who-came-to-dinner is the blue-white light that the Parchment Worm gives off. This blind bottom-dweller lights up like a firefly when it is disturbed. Its light is so bright that the tips of its tubes may glow at night.

The Plumed Worm (Plate 2) constructs a straight parchment tube with a carefully camouflaged doorway. The funnel-shaped entrance which projects above ground is decorated with an assortment of pebbles, broken shells and seaweed. At low tide, the foot-long worm remains below ground. As soon as water covers the flats, its head and scarlet plumed bill's appear in its doorway. A meat-eater, it reaches out with a set of business-like jaws to catch passing animals. When not feeding, it collects bits of shells and cements them to the outside of its tube. In an aquarium a Plumed Worm will use tinsel, cloth and even feathers for this decorating job.

The Plumed Worm's handiwork can be found everywhere on the flats, but it is difficult to see a live animal. At any disturbance from above, the animal retreats to the bottom of its tube, three feet underground. If you swim at half-tide, using a face mask, you may catch a glimpse of the worm's iridescent body and spectacular blood-red gills. When you reach out a hand, however, it will quickly pull back.

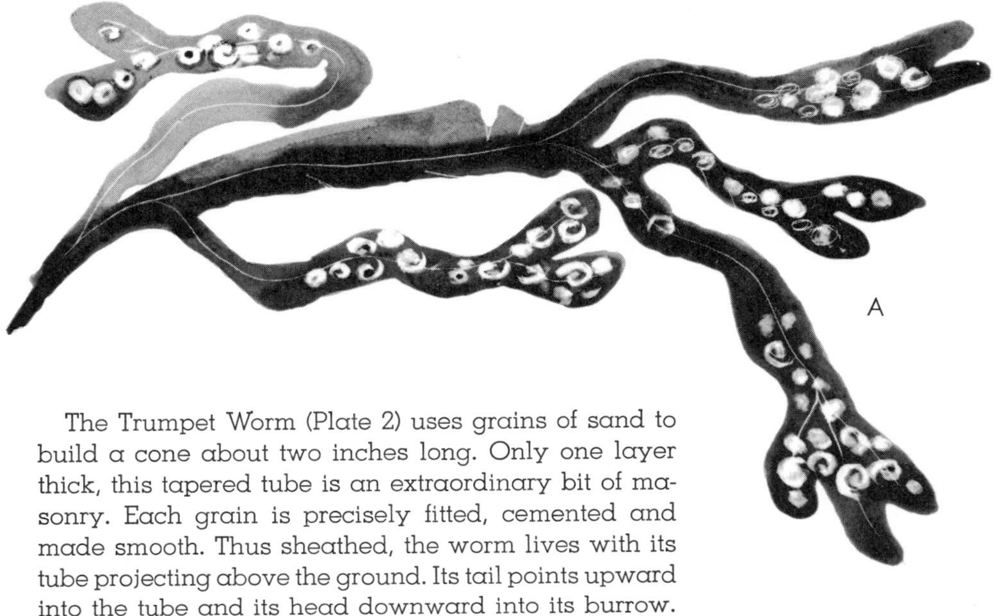

The Trumpet Worm (Plate 2) uses grains of sand to build a cone about two inches long. Only one layer thick, this tapered tube is an extraordinary bit of masonry. Each grain is precisely fitted, cemented and made smooth. Thus sheathed, the worm lives with its tube projecting above the ground. Its tail points upward into the tube and its head downward into its burrow.

An engineer as well as a mason, the Trumpet Worm digs a shaft connecting its burrow with the surface. The shaft keeps filling up and the worm keeps digging with a pair of sharp-pointed golden combs. It feeds on bits of organic matter that slip down the shaft into its burrow, and passes the "processed" sand back up to the surface through its mosaic tube. Although the worm itself is rarely visible, pieces of its tube often wash up on the beach.

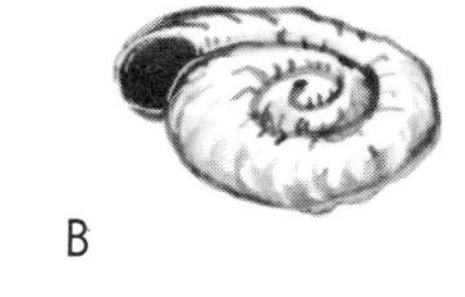

Scarcely anyone notices the most abundant worm of all, a tiny creature that is able to take lime from the water as mollusks do, and build itself a home. Never popular enough to have acquired a common name, *Spirorbis borealis* cements minute coiled tubes to blades of seaweed and the undersides of stones. In order to feed, the worm thrusts out a wreath of feathery tentacles, which trap plankton and act as gills. When the tentacles are pulled back, a cone-shaped plug closes the opening of the tube.

Completely self-sufficient inside its spiral tube, the worm is hermaphroditic. One end is female and the other male. Its eggs are kept in the tube until they are ready to hatch. Twice each month the parent worm

Spirorbis Worm

A. On seaweed

B. Tube (enlarged)

C. Worm (greatly enlarged)

lays a new cluster of eggs and, at the same time, sends a batch of young to sea. The birth of the worms always takes place during a neap tide when the gentler ebb and flow of the water offers these minute animals a better chance to survive. After only an hour's swimming, the young worms start to build tubes. The *Spirorbis* way of life is so successful that if they were not a favorite food of young Purple Snails they might overrun the low-tide world.

Several larger relatives of *Spirorbis* build twisted tubes of lime on clam and oyster shells. Their tubes, up to three inches long, are cemented one on top of the other and every which way, until they sometimes cover a whole shell. One of these tube-builders is *Hydroides dianthus*, another worm with richly-colored graceful tentacles. The biologist who gave the worm the second part of its name thought that its plumed gills looked like pink dianthus flowers.

Tube Worms

Oyster farming

OYSTER BEDS

Three hundred years ago oysters were a poor man's food, to be had for the picking when nothing tastier was at hand. Some were a foot long and so meaty that they had to be cut into pieces before they could be swallowed. They formed such extensive beds along the shores that "The oyster bankes do barre out the bigger ships" one New England settler complained.

American oysters were a source of surprise to Englishmen whose native oysters are much smaller. When novelist William Thackeray was served Cape Cod oysters during a tour of the United States, he said they were so big that he felt as if he had "swallowed the baby."

Today the natural beds are gone, destroyed by overfishing and water pollution. Although "wild" oysters may be found on the flats or on breakwaters nearby, almost all of the oysters of the Outer Lands have been planted there, and are cultivated and harvested by oyster farmers.

Like the clams, oysters are two-shelled mollusks. They thrive in protected bays and in the mouths of rivers where the water is less salty than in the open sea. During the warm months, they release enormous

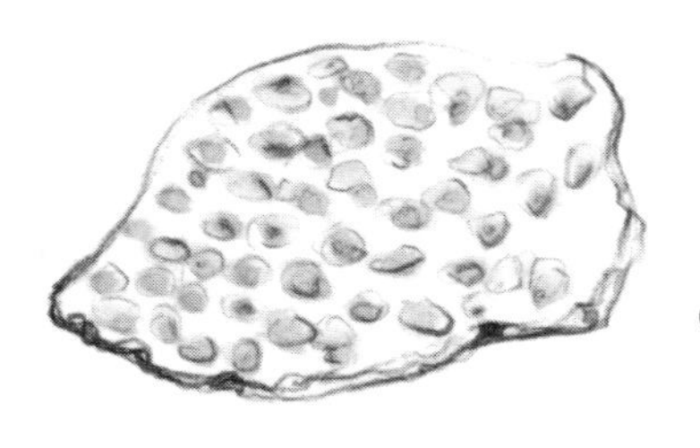

Oyster spat

Oyster, 9 months

numbers of eggs and sperm into the water. A mature female may produce 300 million eggs in one summer—roughly a hundred times as many as a Soft-shell Clam.

After two or three weeks of swimming, the oyster, then about the size of a sand grain, drops to the bottom. With the help of a tiny clamlike foot, it crawls around, searching for a place to settle down. Lacking a siphon, it cannot live underground. Instead it must find a clean hard object to fasten on to, so that it will be raised slightly above the surface. Once it has cemented its bottom shell to a suitable spot, its foot disappears and it remains anchored for life.

Even young oysters—known as spat—feed by opening their shells and straining plankton from the water. Fine hairlike projections wave back and forth, producing a current of water that flows through their gills and out of the shell again. Their pumping mechanism is so efficient that an adult oyster can pump up to forty quarts of water in an hour. You can see this machinery in operation if you visit an oyster bed. As the tide covers it, little fountains of water squirt up everywhere.

1 year

2 years

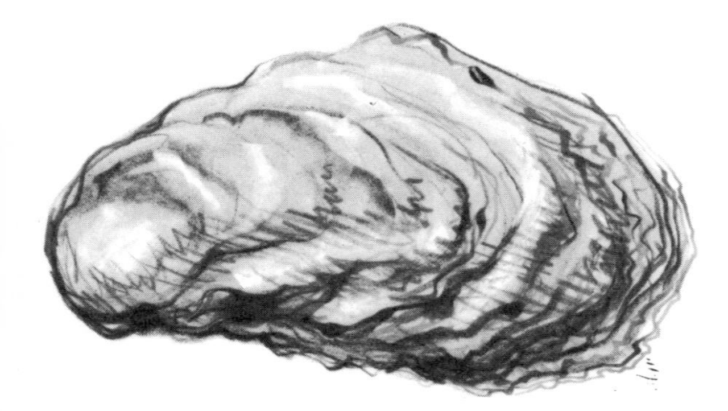

Eastern Oyster
3–4 years

At low tide, or when a shadow which might mean danger falls across its shell, the oyster closes up tight. In deeper water it pumps continuously, for perhaps twenty-two hours out of twenty-four. When water temperatures drop in winter, the animals stop feeding and hibernate.

A mantle similar to that of a clam forms the oyster's shells and adds to them each year. The animal's body lies in the curved bottom shell while the flat top shell opens and closes like the lid of a box. The shells fit so snugly that oysters can live for weeks out of water if they are kept at low temperature.

Occasionally a grain of sand or other foreign object is trapped between the mantle and the shell. Unable to remove it, the mantle coats it with the same smooth material that makes up the inside of the shell. This "mother-of-pearl" lining, produced by most mollusks, consists of thin sheets of calcium carbonate crystals, overlaying a horny material called conchiolin. Layer after layer covers the sand grain until it becomes imbedded in the shell. If it has moved about during this process, so that all sides of it are coated, it ends up as a rounded pearl.

Any oyster—and most mussels and clams—can make pearls, but the lustrous gem pearls come from oysters of a different family that live in tropical waters. Cultured pearls are made by cutting a hole in the Pearl Oyster's mantle and putting a tiny bead inside. The animal does the rest, taking four years to form a pearl of marketable size.

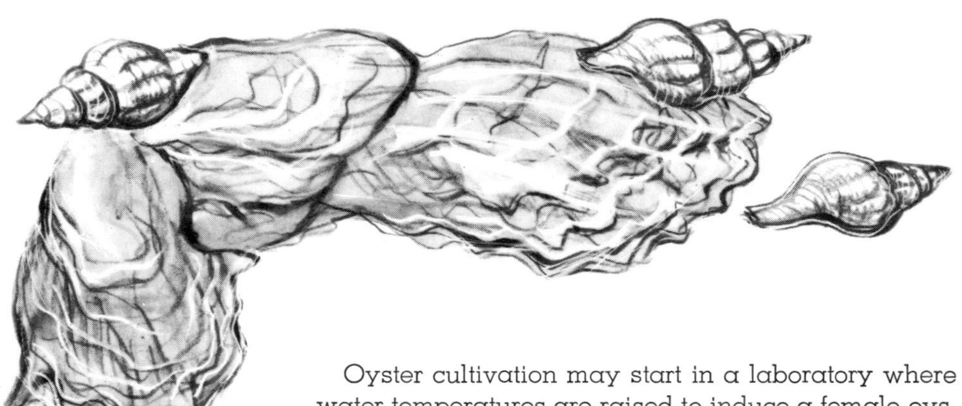

Oyster Drills

Oyster cultivation may start in a laboratory where water temperatures are raised to induce a female oyster to lay eggs. More often it begins outdoors. The oyster farmer clears his ground and puts out old oyster and scallop shells, or man-made collectors, to catch the spat when they are ready to attach. In areas where oysters no longer spawn, the farmer buys "seed" oysters, ranging in size from one-quarter of an inch to two inches, and plants them on his bed. He transplants his crop frequently, moving it to deeper water where more food is available, and separating the clustered shells so that each individual will have room to grow. When the oysters are three or four years old, they are often moved to special fattening grounds where, because of particularly favorable conditions, they grow plump and flavorsome. A "Cape Cod" oyster may have hatched in Chesapeake Bay, spent its youth in Long Island Sound and its last years in Wellfleet or Cotuit harbor.

The custom of eating oysters only during months that have "r" in their names originated in England hundreds of years ago. Before the days of modern refrigeration oysters were likely to spoil during warm weather. Also, English oysters, unlike our own, lay their eggs inside their shells. Eating these oysters during the summer when the eggs are incubating would mean the destruction of future crops. American oysters are sometimes thin and watery after egg-laying, but except for this possible loss of flavor, there is no reason not to eat them twelve months a year.

Oyster Crab

Oyster Drills

Man shares his fondness for oysters with starfish and fish, crabs and snails. One of the worst pests on oyster beds is a handsome snail, the Oyster Drill. It climbs on top of its helpless victim and drills a hole through the shell with its rasplike tongue. After scraping out the meat, the Drill often lays its eggs, enclosed in tan, flask-shaped cases, on the empty shell.

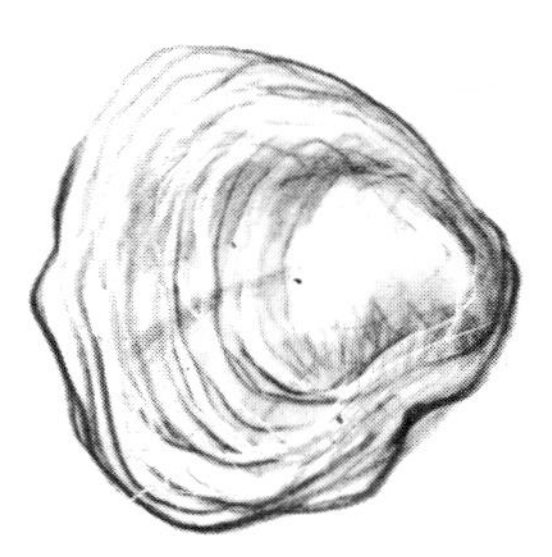

Jingle Shells

Oyster Crabs

In addition to their outright enemies, oysters sometimes play host to Pea Crabs. Oyster Crabs belong to the same family as the Parchment Worm's boarders, but only the female lives with the oyster, while the much smaller male swims in on occasion to mate with her. The crab stations herself on the oyster's gills, taking her pick of the food as it floats by. Although her relationship with her host is described as commensal, the oyster certainly gets the worst of the bargain. Revenge, in the oyster's name, can be sweet. Try sautéing Oyster Crabs in butter and popping them into your mouth.

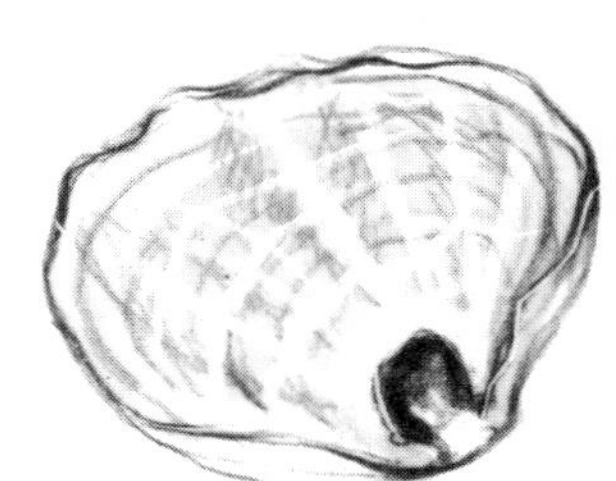

Jingle Shells

When the oyster farmer puts out shells to catch the spat, he creates a new environment along the sandy shores. In waters teeming with minute animals hunt-

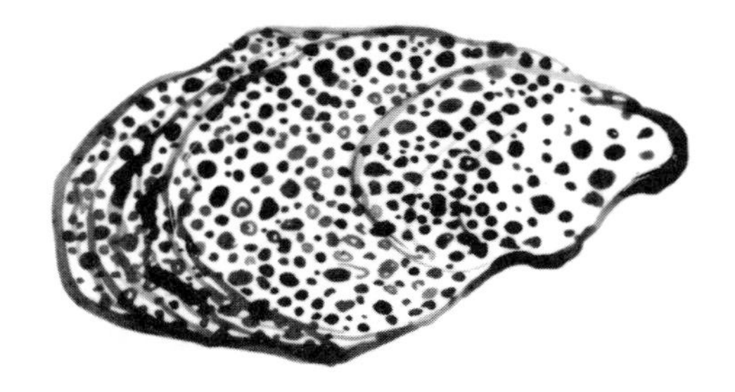

Oyster shell riddled by Boring Sponge

ing for anchorage, the shelly bottom attracts hydroids and Tube Worms, Boat Snails, Jingle Shells and sponges. The fragile gold or silver Jingle Shells are so flat that it is hard to believe they house a living animal. They are bivalves who fasten themselves to shells or stones through a hole in their bottom shell. If large numbers of Jingles settle on an oyster bed, they compete with the oysters for living space. Growing more rapidly than the young oysters, their shells sometimes spread over the spat and smother them. Jingles get their name from the metallic sound that the empty shells make when they rattle against each other on the beach.

Boring Sponges

Several kinds of sponges grow on oyster beds, but only the Boring Sponge causes concern to oyster farmers. Microscopic in size when it settles, the sponge tunnels through living and dead shells. Although no one is sure how this is accomplished, the sponge probably uses an acid to dissolve the limy substance of the shell. Growing slowly inside its tunnels, the sponge riddles the shell with holes and forms a sulphur-yellow crust on its outer surface. Eventually the shell crumbles to bits. Troublesome on oyster beds, Boring Sponges are useful in the larger economy of the sea. By destroying shells that would otherwise accumulate in enormous numbers, they perform the same sort of cleanup job in the water that bacteria and fungi do on land.

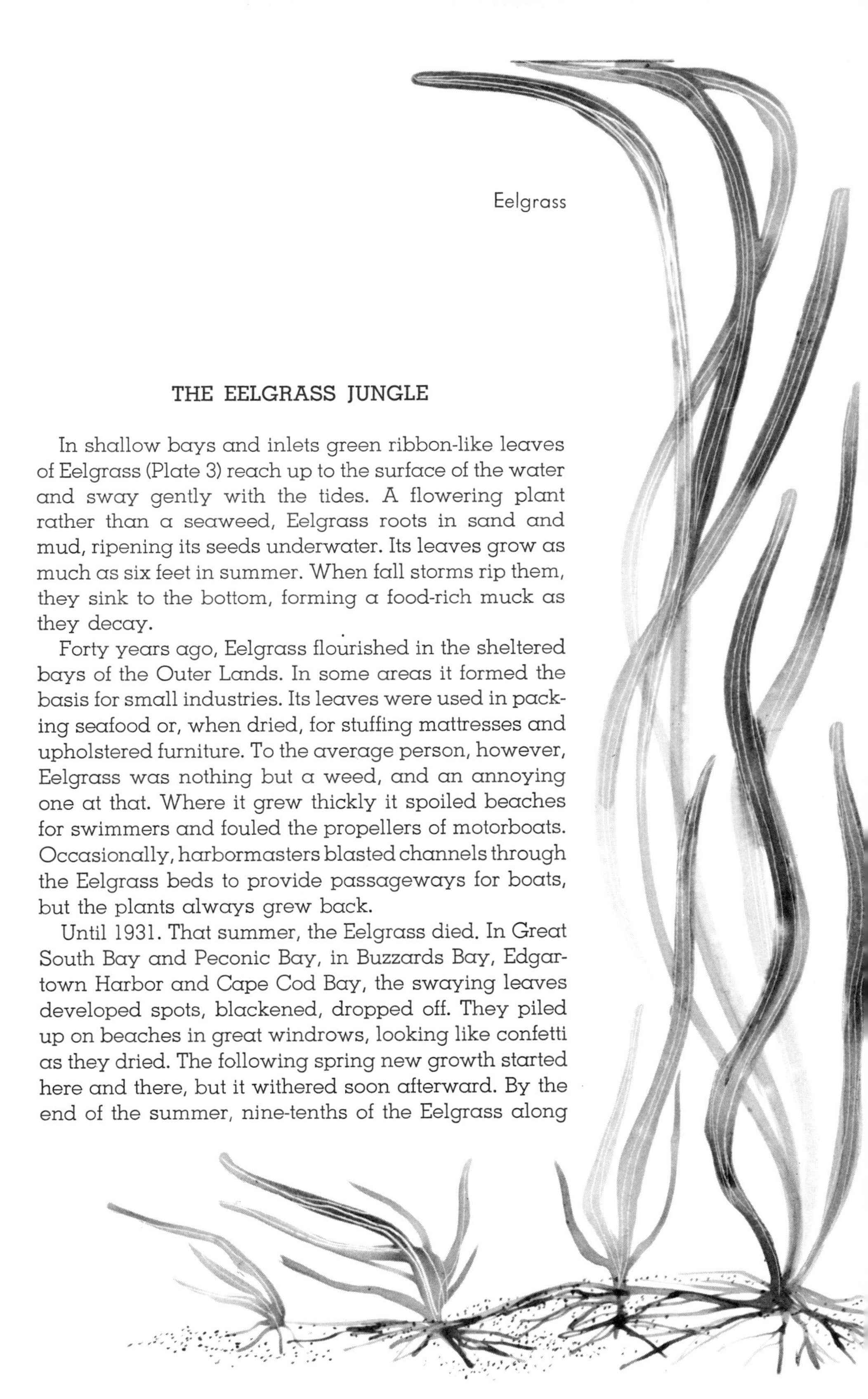
Eelgrass

THE EELGRASS JUNGLE

In shallow bays and inlets green ribbon-like leaves of Eelgrass (Plate 3) reach up to the surface of the water and sway gently with the tides. A flowering plant rather than a seaweed, Eelgrass roots in sand and mud, ripening its seeds underwater. Its leaves grow as much as six feet in summer. When fall storms rip them, they sink to the bottom, forming a food-rich muck as they decay.

Forty years ago, Eelgrass flourished in the sheltered bays of the Outer Lands. In some areas it formed the basis for small industries. Its leaves were used in packing seafood or, when dried, for stuffing mattresses and upholstered furniture. To the average person, however, Eelgrass was nothing but a weed, and an annoying one at that. Where it grew thickly it spoiled beaches for swimmers and fouled the propellers of motorboats. Occasionally, harbormasters blasted channels through the Eelgrass beds to provide passageways for boats, but the plants always grew back.

Until 1931. That summer, the Eelgrass died. In Great South Bay and Peconic Bay, in Buzzards Bay, Edgartown Harbor and Cape Cod Bay, the swaying leaves developed spots, blackened, dropped off. They piled up on beaches in great windrows, looking like confetti as they dried. The following spring new growth started here and there, but it withered soon afterward. By the end of the summer, nine-tenths of the Eelgrass along the coast from the Carolinas to Labrador had disappeared. In Sweden, Holland, France, England, the same thing happened. A pestilence as deadly as the black plague had destroyed the Eelgrass on both sides of the Atlantic.

The Eelgrass story never made newspaper headlines, but while botanists hunted for the killer (finally, although not positively, identified as a waterborne fungus), disturbing reports trickled in to scientific journals. Fishermen were catching less flounder and cod. There were fewer lobsters in the offshore waters. Clams and mussels were disappearing and Bay Scallops were rarely found. Wild ducks and geese migrating along the coast—particularly the Brant Geese who fed almost exclusively on Eelgrass—were starving.

The sudden death of the Eelgrass dramatized its importance. Its creeping roots had anchored the shifting sands. Its waving leaves had trapped silt and sewage that now was smothering clam and oyster beds. And together roots and leaves had provided food, shelter and nursery grounds for a host of fish, shellfish and waterfowl. Many marine animals, including the creatures of the plankton, lived on the decaying plant materials that Eelgrass added to the sandy bottoms of the bays. Larger animals fed on these smaller ones so that the death of the Eelgrass meant the breakup of closely knit seaside communities.

Seed clams were sown on some flats to replace clams that had died. Government officials supplied grain to the migrating waterfowl, and the surviving

French mussel farm

Brants switched to a diet of Sea Lettuce. But large shore areas remained barren wastelands until slowly, during the 1940s and '50s, the Eelgrass began to come back. It is not as lush as it used to be, but wherever it has re-established itself, clumps of mussels are fastened to the matted roots; fish and young eels (Plate 3) weave in and out of the green thickets; and young scallops, Pipefish, and Sea Horses cling to the slippery leaves.

The Eelgrass story may not end happily, however. Because the plants are sensitive to changes in water quality, their comeback is threatened in some areas by the heated water discharged from power plants, in others by pollutants from sewage systems.

Mussels

The Edible Mussel (Plate 3), also called the Blue Mussel because of the rich blue of its shell, is a near relative of oysters and clams. Hatching in the water, the young mussel changes quickly from swimmer to pedestrian. Its long tubular foot does double work, allowing the mussel to crawl around and to anchor to a support. A sticky liquid, pouring out of the mussel's foot, hardens into a thread after minutes in the water. Unlike a young clam which "spins" a single byssus thread, the mussel keeps on "spinning" until its anchor lines stretch in all directions. These frail-looking golden threads can take a surprising amount of heavy surf without breaking. The byssus threads "spun" by Pen Shells are used to weave delicate cloth-of-gold fabrics.

Although mussels rarely move when they grow older, they can cast off if they have to. A mussel climbs up a steep slope in much the same way a mountain climber does—by fastening its threads higher and higher and slowly inching up to them.

Edible Mussels live on the surface, feeding on plankton that their gills strain from the water. In Eelgrass thickets they fasten to their neighbors' shells, but in other areas they anchor to rocks or wharf pilings. Europeans cultivate mussels by "planting" trees or floating rafts in shallow water so the young animals will have places to attach to. Mussel farming is just beginning to be talked about in the United States.

Scallops

The Bay Scallop's (Plate 3) fluted shell was once the badge of pilgrims on their way to the Holy Land. Now an oil company has taken it as its trademark, emblazoning it on billboards around the world. The animal itself is the liveliest of all our bivalves. Although it fastens to the Eelgrass with byssal threads, it also swims, rapidly and noisily, by opening and closing its shells. As the shells snap shut, a stream of water is forced out and the jet-propelled scallop shoots ahead.

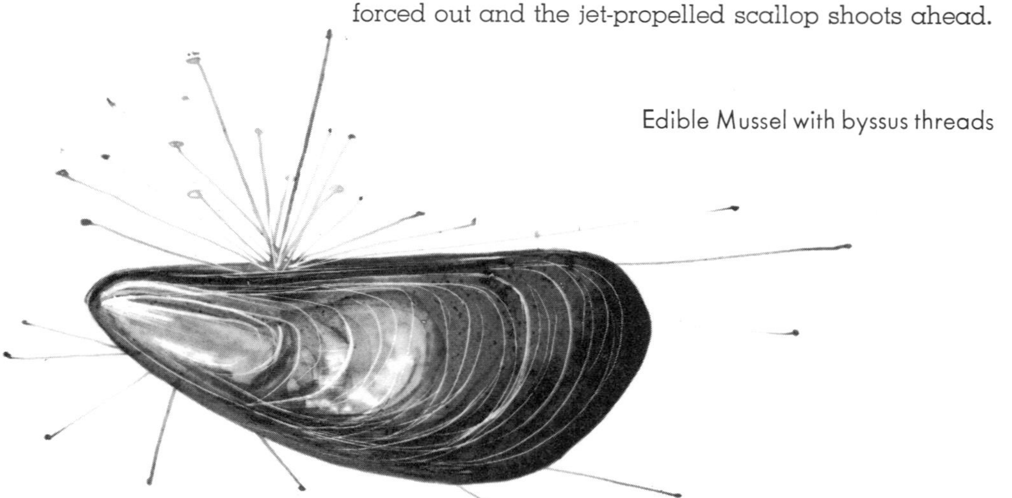

Edible Mussel with byssus threads

Sea Scallop (reduced)

Bay Scallop

Able to control the direction of the jet of water, the animal also zooms backward and sideways, its shells clacking like castanets.

As befits a traveler, the scallop's nervous system is more complicated than that of its sedentary neighbors. Fringed with tentacles, its mantle is also decorated with thirty to forty bright blue eyes. These are working eyes, each with its own cornea, lens and retina. Although the scallop probably cannot form an image with them, it recognizes changes in light intensity. The animal also possesses a chemical sense which permits it to "taste" the approach of a starfish or Skate and take off in the opposite direction.

The mobility of Bay Scallops gives them a better chance to find food. Although they are too lightweight to swim against the current, within the sheltering Eelgrass they can raise themselves above the ground to take in plankton-rich water. Growing faster than oysters or clams, they live for a much shorter time. They spawn when they are a year old and die of old age about the age of two.

As great numbers of scallops die in early summer, their shells drift to shore. On the back of each shell you can see distinct annual growth lines. These ridges are formed in the spring when the animal starts to grow again after a winter's hibernation. Notice also the wavy edge of the shell which long ago added the word "scalloped" to our vocabularies.

Scallops have been harvested only in the last hundred years. Before that time, perhaps because they have bright orange and yellow bodies as well as blue eyes, they were believed to be poisonous. Even today only the muscle that opens and closes the shell (inaccurately called the "eye") is marketed, while the rest is thrown away. Despite old-timers' reports of cats whose tails and ears dropped off after a scallop dinner, the entire scallop is edible—and delicious.

Sea Scallops are giant cousins of the Bay Scallop. Living in deeper colder waters, they are brought to the surface by dredging. Although some Sea Scallops are found in Cape Cod Bay, most of the catch comes from Georges Bank, far from shore.

Starfish and Brittle Stars

In the Eelgrass community, starfish and mussels go together. Like cat and mouse or wolf and sheep, they are linked by stern jungle law. At low tide starfish of all sizes and colors sprawl on the flattened Eelgrass leaves. Piled one on top of the other, they lie as limp and motionless as the melting watches in a painting by Salvador Dali. Even when the tide rises and they begin to move, their behavior has a dreamlike quality.

Starfish regenerating

More properly called sea stars since they are no kin to fish, starfish are classified as echinoderms, a word that means "spiny-skinned." Usually they have five arms radiating from a central disk. Water and food enter this central region, the water through a stony sieve on the upper surface and the food through a mouth underneath. Flowing through canals to the starfish's arms, the water helps to operate its tube feet.

Each starfish has rows of hollow feet on its underside. Suckers at the tips of these feet cling tightly to a surface, while changes in water pressure force the feet to lengthen and contract. The starfish moves in a 1-2-3 rhythm, stretching out its feet, attaching the suckers, and then contracting the tiny tubes. As the feet grow shorter, they slowly but surely pull their owner ahead.

But which way is ahead? Since a starfish has neither head nor tail, any arm may lead the way, with the others following. Tube feet at the end of each arm seem to act as sense organs. The exact way that these feelers work is unclear. However, when they signal "Mussel on the port side!" there's no need for the animal to turn. Instead the port arm takes over until the animal finds its prey.

Starfish moving

Starfish

Fastening its feet to both of the mussel's shells, the starfish humps up its body and begins to pull. The mussel pulls too, holding its shells tightly closed. At first the tug-of-war is a draw, but the starfish, using its feet in relays, has time on its side. Wearied by the struggle, the mussel at last relaxes and its shells slowly part.

The starfish now faces another problem. Lacking claws, it has no way of moving the mussel to its stomach—so it moves its stomach to the mussel. Literally. Turned inside out, the thin baglike stomach of the starfish slides into the mussel shell. Pouring out digestive juices, it eats the mussel then and there. After the shell has been emptied, the stomach is stowed away inside the starfish again.

Most starfish lay their eggs in the water during the summer. The young attach themselves to rocks or Eelgrass for a time, starting to feed on baby mussels and clams even before their arms are fully formed. Starfish can go for months without food, but when it is abundant they grow rapidly.

The Common Starfish (Plates 3 and 4) is sometimes eleven inches in diameter while the Purple Star (Plate 4) may reach seventeen inches when living in cold Maine waters. The colors of both of these sea stars vary greatly, ranging from brown to purple, green or orange. The best way to tell them apart is to look at the stony plates through which they take in water. The Common Starfish's sieve is bright orange while the Purple Star's is a pale yellow.

Brittle Stars (Plate 4) also live in the Eelgrass. Lacking suction cups on their feet, they move by wriggling their long flexible arms. They take shelter in a clump of mussels, twisting in and out of the golden byssus threads as they hunt for worms and other small creatures. Blood Sea Stars (Plate 4) are found in rock crevices and tide pools. They are cool-water animals, usually living offshore south of Cape Cod.

Starfish are second only to man in their destructiveness. One star has been known to eat seven yearling oysters in a day. When large numbers of them attack an oyster bed they can destroy it in short order. Their lone saving grace as far as oystermen are concerned is that they cannot live in water with a low salt content. For this reason, they are seldom found on oyster or mussel beds near the mouths of streams, where the water is fresher.

Oyster farmers now use chemical and mechanical controls to get rid of starfish, but years ago they simply cut them in two with a hatchet. If particularly annoyed by the invaders, they might chop them into three or four pieces. After several months, they had twice, three times or four times as many starfish as before—because starfish are able to regenerate. They can grow new arms if they lose them. Experiments with the Purple Star have shown that a single arm attached to only one-fifth of its central disk can form a whole new starfish. Brittle Stars whose skinny arms are easily broken off can also regenerate the missing parts.

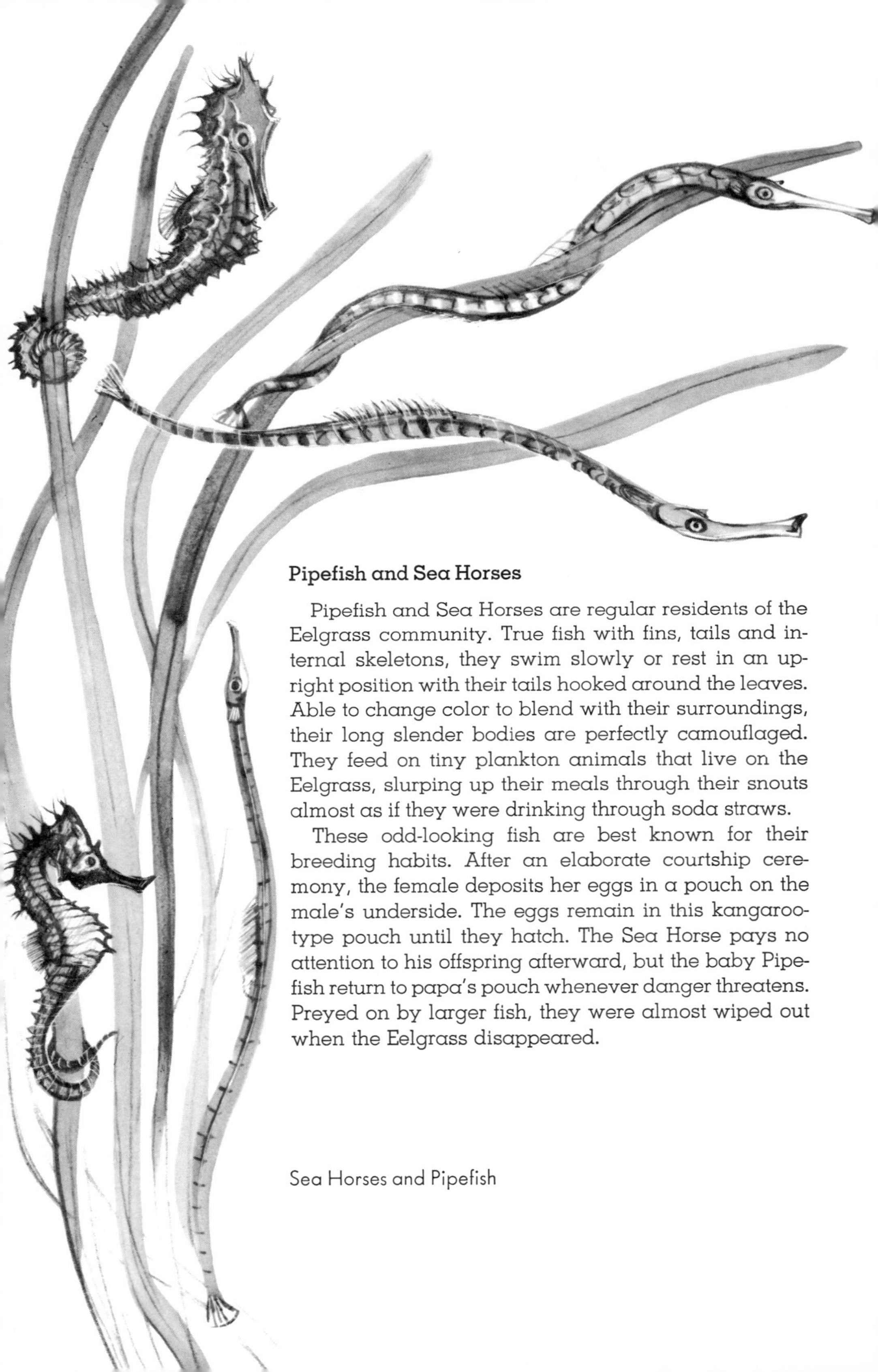

Pipefish and Sea Horses

Pipefish and Sea Horses are regular residents of the Eelgrass community. True fish with fins, tails and internal skeletons, they swim slowly or rest in an upright position with their tails hooked around the leaves. Able to change color to blend with their surroundings, their long slender bodies are perfectly camouflaged. They feed on tiny plankton animals that live on the Eelgrass, slurping up their meals through their snouts almost as if they were drinking through soda straws.

These odd-looking fish are best known for their breeding habits. After an elaborate courtship ceremony, the female deposits her eggs in a pouch on the male's underside. The eggs remain in this kangaroo-type pouch until they hatch. The Sea Horse pays no attention to his offspring afterward, but the baby Pipefish return to papa's pouch whenever danger threatens. Preyed on by larger fish, they were almost wiped out when the Eelgrass disappeared.

Sea Horses and Pipefish

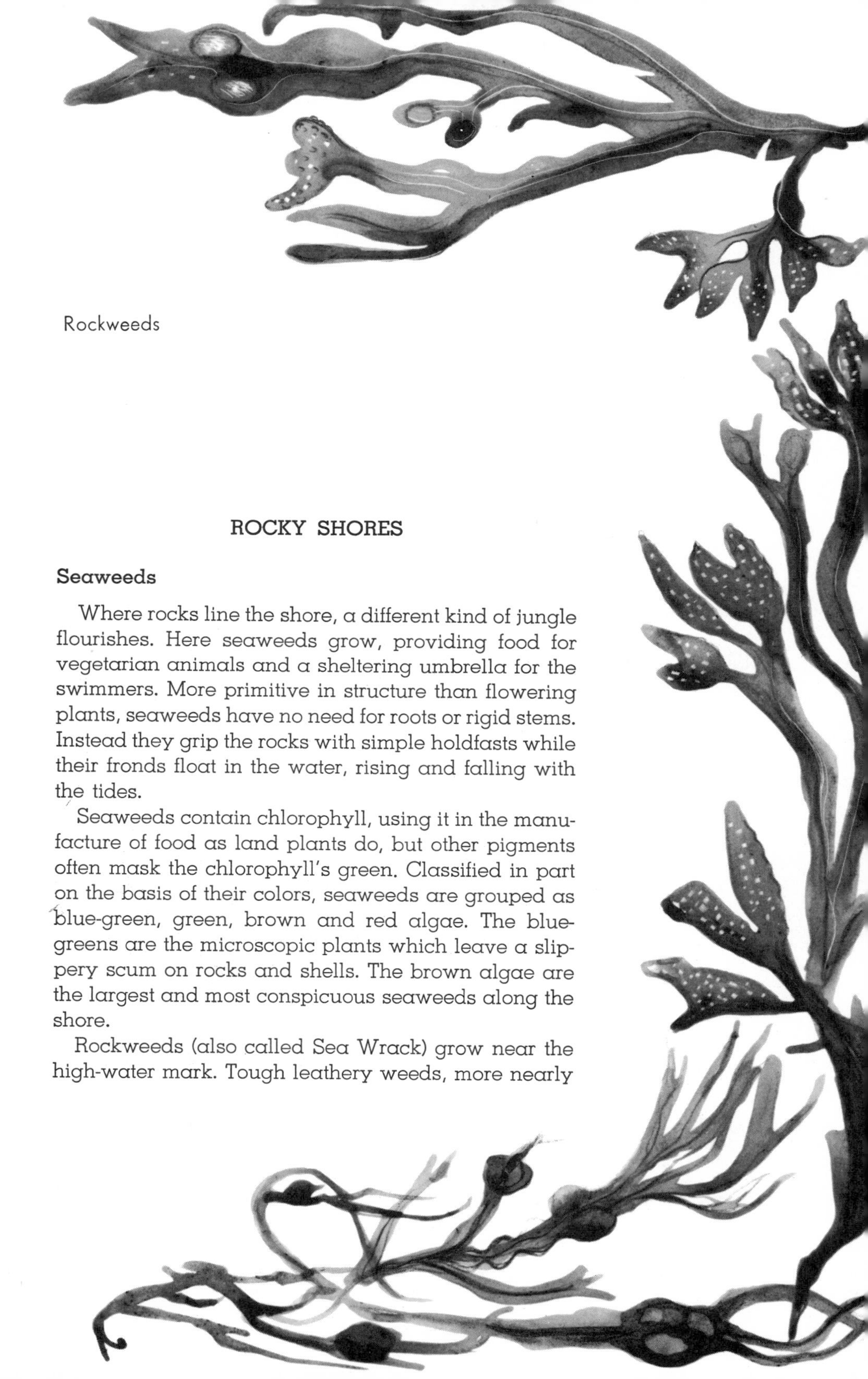
Rockweeds

ROCKY SHORES

Seaweeds

Where rocks line the shore, a different kind of jungle flourishes. Here seaweeds grow, providing food for vegetarian animals and a sheltering umbrella for the swimmers. More primitive in structure than flowering plants, seaweeds have no need for roots or rigid stems. Instead they grip the rocks with simple holdfasts while their fronds float in the water, rising and falling with the tides.

Seaweeds contain chlorophyll, using it in the manufacture of food as land plants do, but other pigments often mask the chlorophyll's green. Classified in part on the basis of their colors, seaweeds are grouped as blue-green, green, brown and red algae. The blue-greens are the microscopic plants which leave a slippery scum on rocks and shells. The brown algae are the largest and most conspicuous seaweeds along the shore.

Rockweeds (also called Sea Wrack) grow near the high-water mark. Tough leathery weeds, more nearly

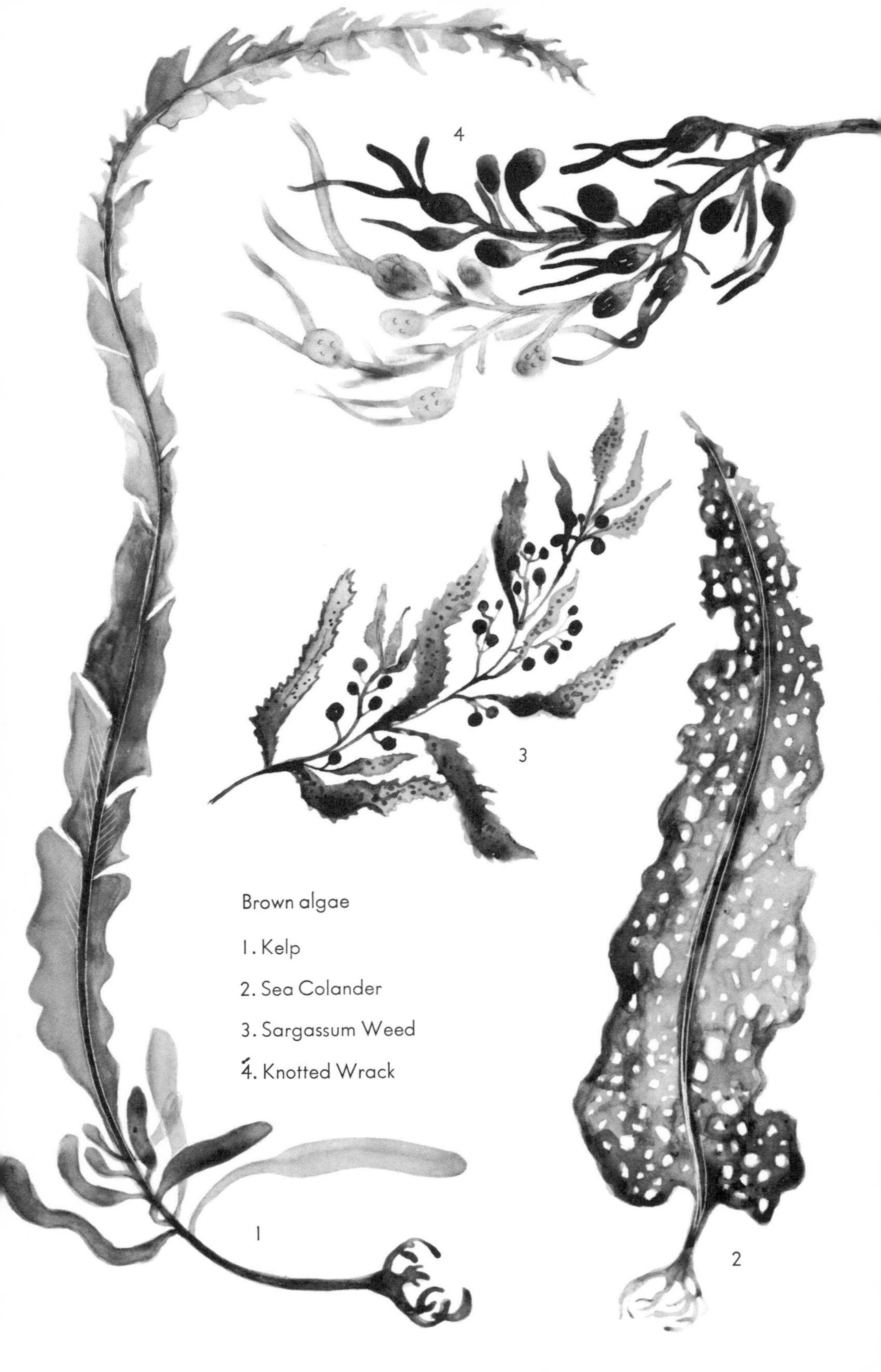

Brown algae

1. Kelp

2. Sea Colander

3. Sargassum Weed

4. Knotted Wrack

olive-green than brown, they become crisp and black when exposed to the air at low tide. Almost all of the Rockweeds have gas-filled bladders on their fronds. These little balloons keep the plants floating near the surface—and make a satisfying "Pop!" when they are pinched.

The more delicate red algae grow lower down. The flat fronds of Dulse (Plate 5), Chenille Weed (Plate 5) which looks as if it is made of red wool, and the closely cropped Irish Moss (Plate 5) need to be covered by water most of the time. Irish Moss, ranging in color from spring-green to purple, turns white when it bleaches in the sun. Used in the manufacture of jellies and puddings, it was once the basis of a considerable industry on Block Island.

Below the meadows of Irish Moss, Coralline algae stain the rocks pink. These unusual plants are able to take lime from the sea as animals do and incorporate it into their tissues. Some of the Corallines form flat old-rose patches on the rocks while others are branching seaweeds with stony pink and white fronds. In tropical waters, Coralline seaweeds play a part in building coral reefs.

Kelps, another group of brown algae, live at the base of the rocks where they are uncovered only by the lowest spring tides. Some kelps grow in deeper water offshore, wrapping their branching holdfasts around clumps of shells and stones. Their long rubbery fronds, which feel like inner tubes, are almost impossible to

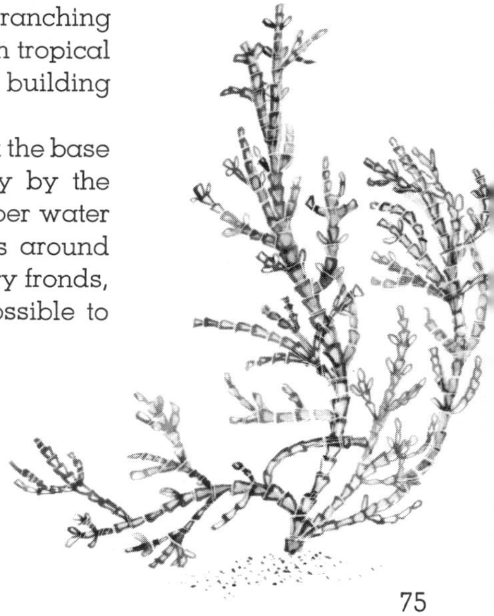

Coralline algae

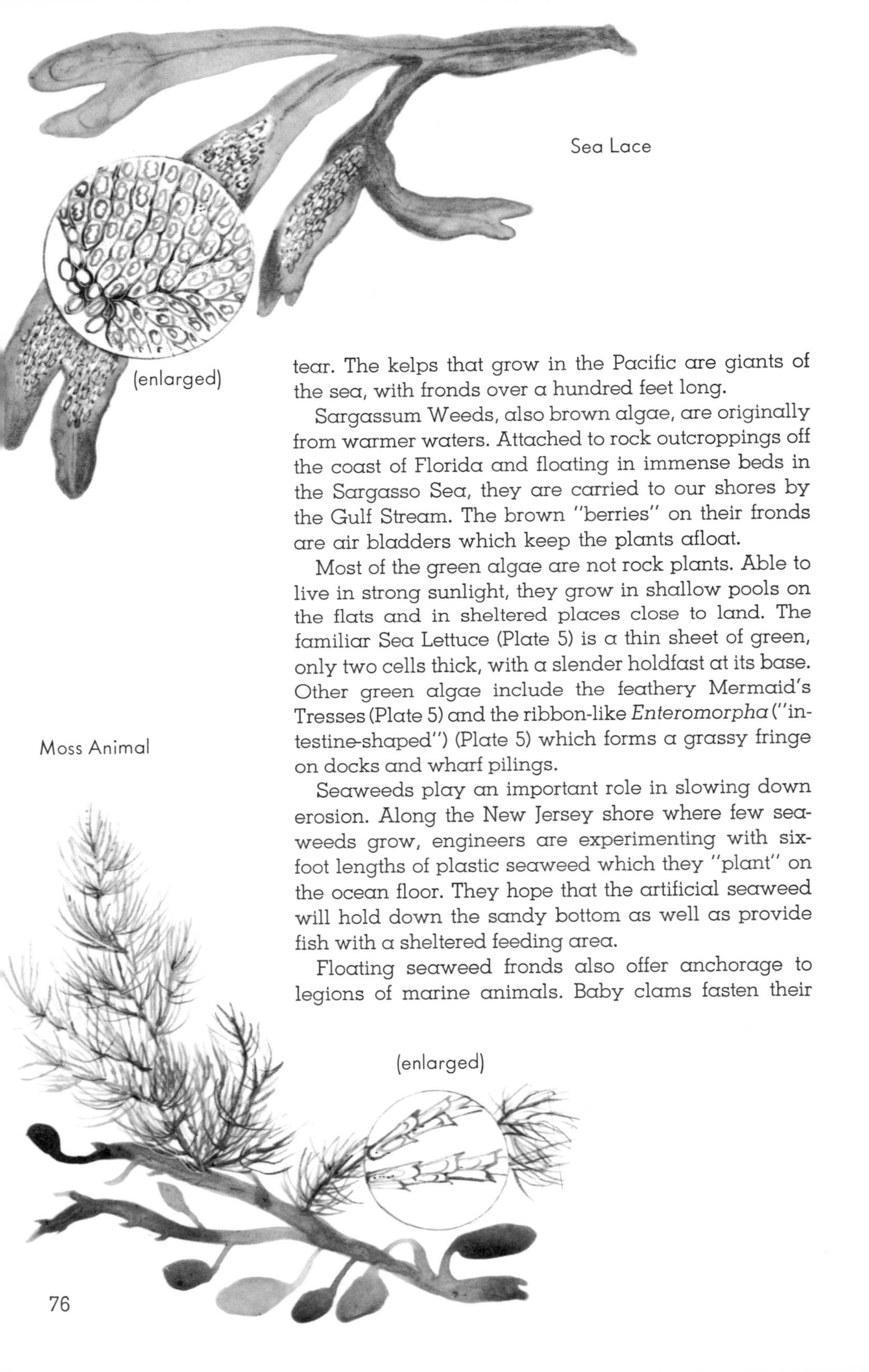

tear. The kelps that grow in the Pacific are giants of the sea, with fronds over a hundred feet long.

Sargassum Weeds, also brown algae, are originally from warmer waters. Attached to rock outcroppings off the coast of Florida and floating in immense beds in the Sargasso Sea, they are carried to our shores by the Gulf Stream. The brown "berries" on their fronds are air bladders which keep the plants afloat.

Most of the green algae are not rock plants. Able to live in strong sunlight, they grow in shallow pools on the flats and in sheltered places close to land. The familiar Sea Lettuce (Plate 5) is a thin sheet of green, only two cells thick, with a slender holdfast at its base. Other green algae include the feathery Mermaid's Tresses (Plate 5) and the ribbon-like *Enteromorpha* ("intestine-shaped") (Plate 5) which forms a grassy fringe on docks and wharf pilings.

Seaweeds play an important role in slowing down erosion. Along the New Jersey shore where few seaweeds grow, engineers are experimenting with six-foot lengths of plastic seaweed which they "plant" on the ocean floor. They hope that the artificial seaweed will hold down the sandy bottom as well as provide fish with a sheltered feeding area.

Floating seaweed fronds also offer anchorage to legions of marine animals. Baby clams fasten their

Pale Periwinkle

silky threads to Sea Lettuce. Rockweeds are covered with the chalk-white tubes of *Spirorbis* worms and the lacy mats of Moss Animals.

Moss Animals

Moss Animals (whose scientific name is Bryozoa) form a crust on the blades and stalks of the weeds. With a magnifying glass you can see that this mosaic is made up of a network of separate compartments, arranged in rows. Each cell is occupied by a minute animal, no more than one-sixty-fourth of an inch long, which reaches out with a circle of tentacles to capture still smaller plankton creatures. Individual Moss Animals live for only a few weeks, but the colony continues to grow, covering a seaweed blade with thousands of limy cells.

In addition to the decorative Sea Laces, other Moss Animals form upright treelike colonies. The largest of these, which may be a foot long, are sometimes mistaken for plants. They grow on shells and pilings as well as on seaweeds.

Rough Periwinkle

Rock Snails and Chitons

While Moss Animals hunt in the sea, snails graze in the seaweed gardens. Periwinkles, the most abundant snails on the rocks, are vegetarians, using their toothed tongues to scrape off layers of plant cells. Rough Periwinkles browse on the exposed upper sur-

Edible Periwinkle

Limpets

faces of the rocks where they are covered by the sea only during spring tides. They are so close to becoming land animals that their gills work as lungs do, taking oxygen from the air rather than the water. They can live out of water for as long as a month—and if they are submerged for too long they will drown.

Pale Periwinkles live in the Rockweed zone, their rounded shells resembling the air bladders on the weeds. Able to remain out of water for only a short time, they hide under wet seaweed when the tide is low. The big Edible Periwinkle has a wider range, commuting between the seaweed and the bare rocks above. It is often called the Common or Shore Periwinkle. At low tide it glues its shell to a rock with a sticky film before closing its operculum.

Interestingly enough, all three of these snails have different breeding habits. The eggs of the land-bound Rough Periwinkle are fertilized internally and remain within the female's body. Her offspring are miniature snails, complete with shells, from the moment they are born. The Pale Periwinkles remain creatures of the Rockweed, fastening their eggs, enclosed in a stiff jelly, to its fronds. Their babies are also fully formed when they hatch. Only the Edible Periwinkles shed eggs and sperm into the water. Hatching in the sea, their young go through a swimming stage before settling down on shore.

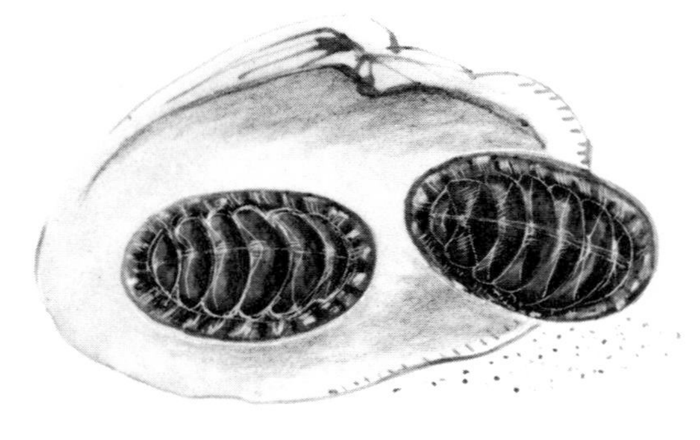

Chitons

Although the Edible Periwinkle's method of spawning sounds hazardous, enormous numbers of these snails cover the rocks. Originally they were not natives of our shores. About a century ago a few of them appeared in Nova Scotia. Multiplying rapidly, they have migrated as far south as Chesapeake Bay and are still going strong. Although seldom eaten in the United States, they are a popular food in England and France.

Limpets and Chitons also live on the rocks. Chitons are primitive mollusks whose jointed shells consist of eight overlapping plates. The flexible plates permit the animal's body to hug the uneven surface of a rock, lying so flat that it is scarcely noticeable. If you pull one off, it will roll up in a ball. Chitons move slowly, feeding on algae that they scrape up with their long tongues. Along the North Atlantic coast, they are seldom an inch long, but a giant western Chiton grows to be a foot long and six inches wide.

Limpets, known to shell collectors as Chinese Hats, are snails with tent-shaped instead of spiral shells. Because a Limpet lacks an operculum, the edges of its shell must fit snugly against a rock so that it can retain water at low tide. The Limpet accomplishes this by shuffling back and forth, grinding down both rock and shell until they form a perfect seal. At low tide and in stormy weather, it clings to the rock with its foot. The stronger the wave, the tighter the animal clings. When a Limpet is really trying, nothing less than a seventy-pound pull can dislodge it.

A Limpet creeps around at high tide, but when the

Purple Snails

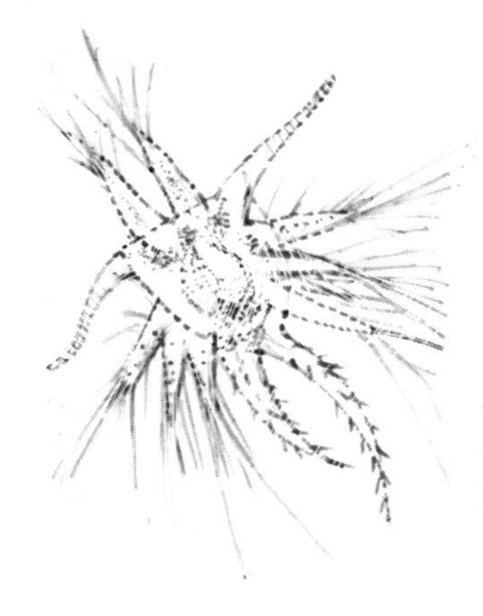

Young barnacle (greatly enlarged)

water ebbs, it always returns to the circular depression that it has gouged out. This is its "home." Experiments have shown that Limpets have some sort of "homing instinct" just as higher animals do.

Thick-shelled Purple Snails live in the Rockweed and in swaying beds of Dulse and Irish Moss. Their name comes not from their shell color—which can be white, yellow or deep brown—but because they secrete a purple substance which was formerly used as a dye. Related snails in the Mediterranean provided the ancient Phoenicians with their famous Tyrian purple.

The Purples lay their eggs on the rocks in parchment cases that turn purple as the eggs develop. From the moment that the baby snails are born they are carnivorous. As tiny creatures they drill through the coiled tubes of *Spirorbis* worms. When they grow larger they feed on mussels, barnacles and other snails. This varied diet affects the color of their shells. A snail that eats barnacles usually has a white shell, while a mussel-eater's shell is chocolate brown. If the animal switches from mussels to barnacles and back again, it ends up with stripes.

Acorn Barnacles

Gooseneck Barnacles

Barnacles

Louis Agassiz, one of the greats of nineteenth-century science, described a barnacle as a "shrimplike animal, standing on its head in a limestone house and kicking food into its mouth with its feet." Long believed to be a mollusk because of its heavy shell, a barnacle is actually a crustacean. For the first three months of its life it is a microscopic swimmer, its changing form resembling the early stages of shrimps and crabs. Then it settles down, cementing itself to a firm substance—which may be a rock, shell or even the thick hide of a whale. In less than a day it constructs a six-sided shell with a hinged door on top. When this operculum opens, the barnacle stretches out its fringed feet to catch food. It closes its doors at low tide, remaining comfortably damp inside.

Barnacles will readily perform in an aquarium or in a jar filled with sea water. Even without a magnifying glass you can watch the quick movements of the animal's feet as they sweep through the water.

The barnacle's home is a permanent one. Within it, the animal grows as crabs do, by shedding its skin and then increasing in size. From time to time it enlarges its limestone cone, probably by dissolving the inner walls and at the same time adding to the plates outside. A hermaphrodite, the barnacle keeps its eggs inside its shell until the young hatch.

The number of barnacles is astronomical. Along a

"Goose Tree"

thousand-yard stretch of rocky shore, a British scientist estimated that there were a billion barnacles, producing up to a trillion young each year. Some fish feed on them at high tide, but their chief enemies are starfish and snails. A Purple Snail can force open a barnacle's operculum and quickly clean out its shell. The empty cone remains on the rock, offering a hiding place for other small sea animals.

The Acorn Barnacle is the familiar barnacle of our shores, but clusters of Gooseneck Barnacles occasionally wash up on the beach. These strange-looking stalked creatures cement themselves to floating objects —logs, empty bottles, buoys—and drift through the water. The muscular leathery stalk contains the animal's ovary, while the major part of its body is enclosed in a five-sided shell. Medieval naturalists who found these barnacles growing on logs believed that they were the fruit of "goose trees" which ripened and hatched into geese. As late as the seventeenth century an imaginative botanist drew a picture of one of these "trees" showing feathered geese emerging from the barnacle shells.

Lobsters

Lobsters lurk underneath the rocks, hidden behind curtains of seaweed. The largest of our invertebrates, the American Lobster lives along the coast from Labrador to North Carolina. In summer it often migrates to shallow water, traveling back to deep water in the fall.

A live lobster has a dark blue-green shell, the blue pigment being changed by heat to the familiar lobster red. Made up of a horny material called chitin and strengthened by lime, the shell is really a jointed outer skeleton. A single sheet of armor covers the animal's back, linking head and thorax. Its business-like claws and four pairs of walking legs grow from the thorax. These details place a lobster as an arthropod (jointed-legged); a crustacean (having a "crusty" shell); and a decapod (ten-legged). Its pinching jaws are considered modified legs.

Each leg has a gill at its base so that as the animal walks a current of water travels over it. The rear part of the lobster's body, its abdomen, is covered with thinner flexible plates which end in a flattened tail-

American Lobster

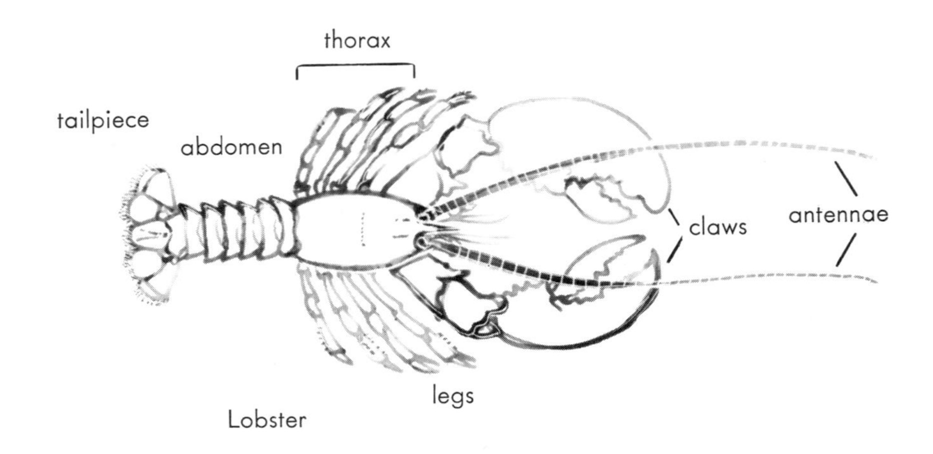

piece. Oarlike swimmerets on the underside of the abdomen carry the lobster forward in the water. If danger threatens, it darts backward by bending and straightening its muscular "tail."

A lobster sees the world, dimly perhaps, through big compound eyes. It explores it with its two pairs of feelers and with thousands of sensory bristles that grow on its armor-plated body. It conquers it with its enormous claws. Holding them forward like a boxer with his guard up, the lobster is ready for all comers. The two claws are not identical. The smaller sharp-toothed claw seizes a fish and cuts it up while the larger one is used to crush clam or mussel shells. A beltline of jaws further breaks up the animal's food and passes it along to its mouth.

More adept at walking than at swimming, the lobster spends its days in rocky caverns or in burrows that it digs in the sand. At night it prowls around, its whiplike antennae sweeping through the water to pick up news of food. It will eat dead fish as well as live ones, Eelgrass, seaweed—and other lobsters.

As it eats and grows, its horny shell becomes too tight. The shell splits lengthwise along the back and at the joint between thorax and abdomen. Lying on its side, the lobster pulls itself out of the cracked shell. A new larger shell has been forming underneath the old one, but it is still soft. For a week or two, the lobster hides in its burrow while the shell hardens. During this time it has a particular craving for lime, eating snails and other mollusks, shell and all.

Molting continues throughout a lobster's life, although less frequently as the animal grows older. A one-pound lobster has probably molted twenty-five times and is about five years old. The largest lobster on record is a forty-two-pounder which may have lived for a century. Lobsters are also able to replace claws and legs if they are lost. At each molt the stump of the missing leg grows larger until, after three or four molts, it is as good as new.

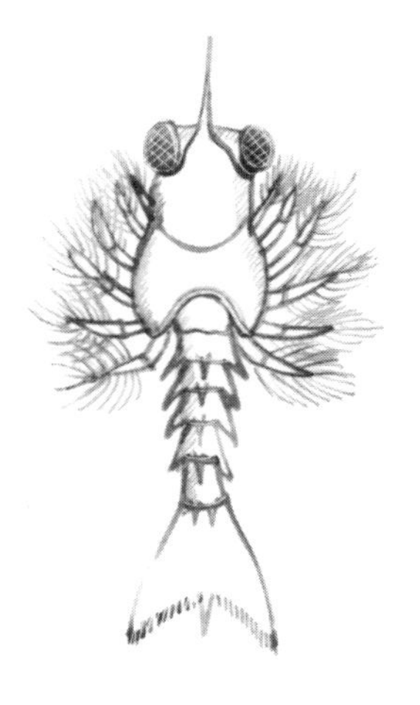

Lobster eggs are more carefully tended than those of almost any other shore animal. The red "coral" found in cooked lobsters is a cluster of unfertilized eggs. After mating, the female moves the eggs to the underside of her abdomen. She incubates them for almost a year, curling under her tail to protect them from enemies and fanning them with her swimmerets to give them oxygen.

When they hatch, young lobsters live the lives of plankton animals. Soft-bodied and transparent, with

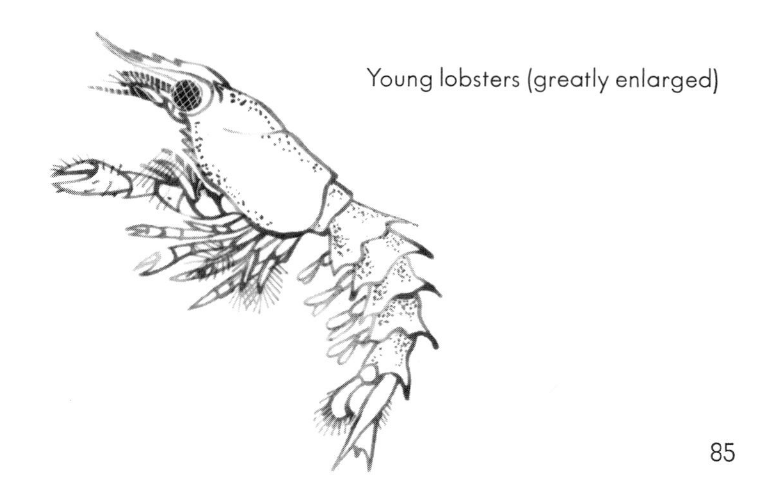

Young lobsters (greatly enlarged)

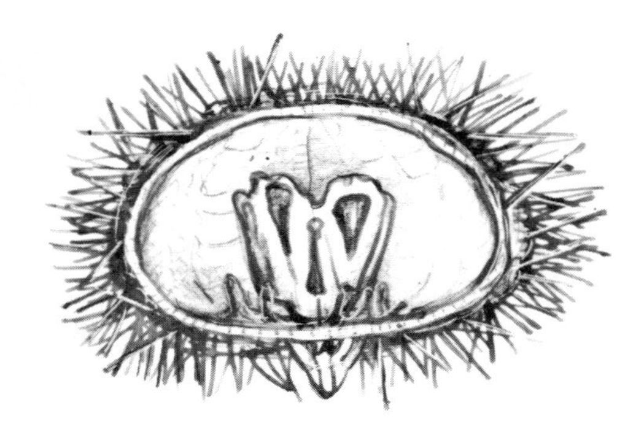

Sea urchin cross-section showing Aristotle's Lantern

round emerald-green eyes, they swim near the surface, gobbling up small creatures and falling prey to large ones. As they molt their appearance changes until they begin to resemble their parents.* Toward the end of their first summer, when they are about an inch long, they sink to the bottom. Too small to defend themselves against cod, Skates and other bottom-feeding fish, they spend most of their time concealed under stones. They are two-year-olds before they venture forth, claws at the ready, to explore.

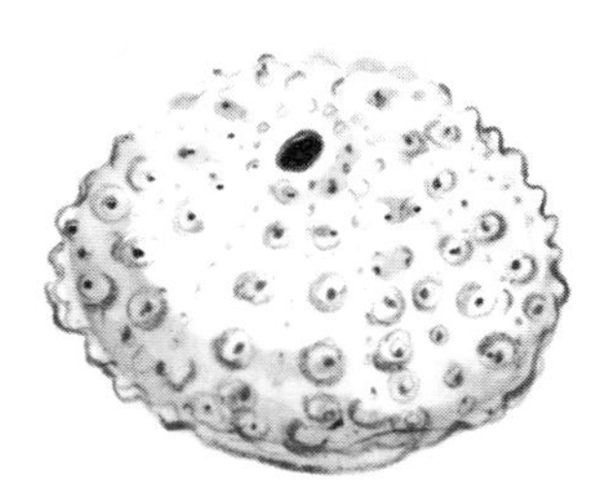

Sea urchin shell

Sea Urchins and Sand Dollars

Sea urchins live in crevices in the rocks or hide underneath them. These prickly animals, whose name comes from an old English word for hedgehog, are related to starfish. Their spines conceal a rounded shell (called a test), which covers the animal's body. Tube feet working like a starfish's poke out through perforations in the shell, and are pulled back when the animal is out of water.

On the shell of a dead urchin, the holes for its tube feet form a striking star-shaped pattern. The pinholes near the top of the shell are openings through which eggs and sperm are shed, while the large hole on the underside is the opening for its mouth. Five sharp

*The Massachusetts Division of Marine Fisheries maintains a lobster hatchery on Martha's Vineyard where lobsters in every stage of development can be seen.

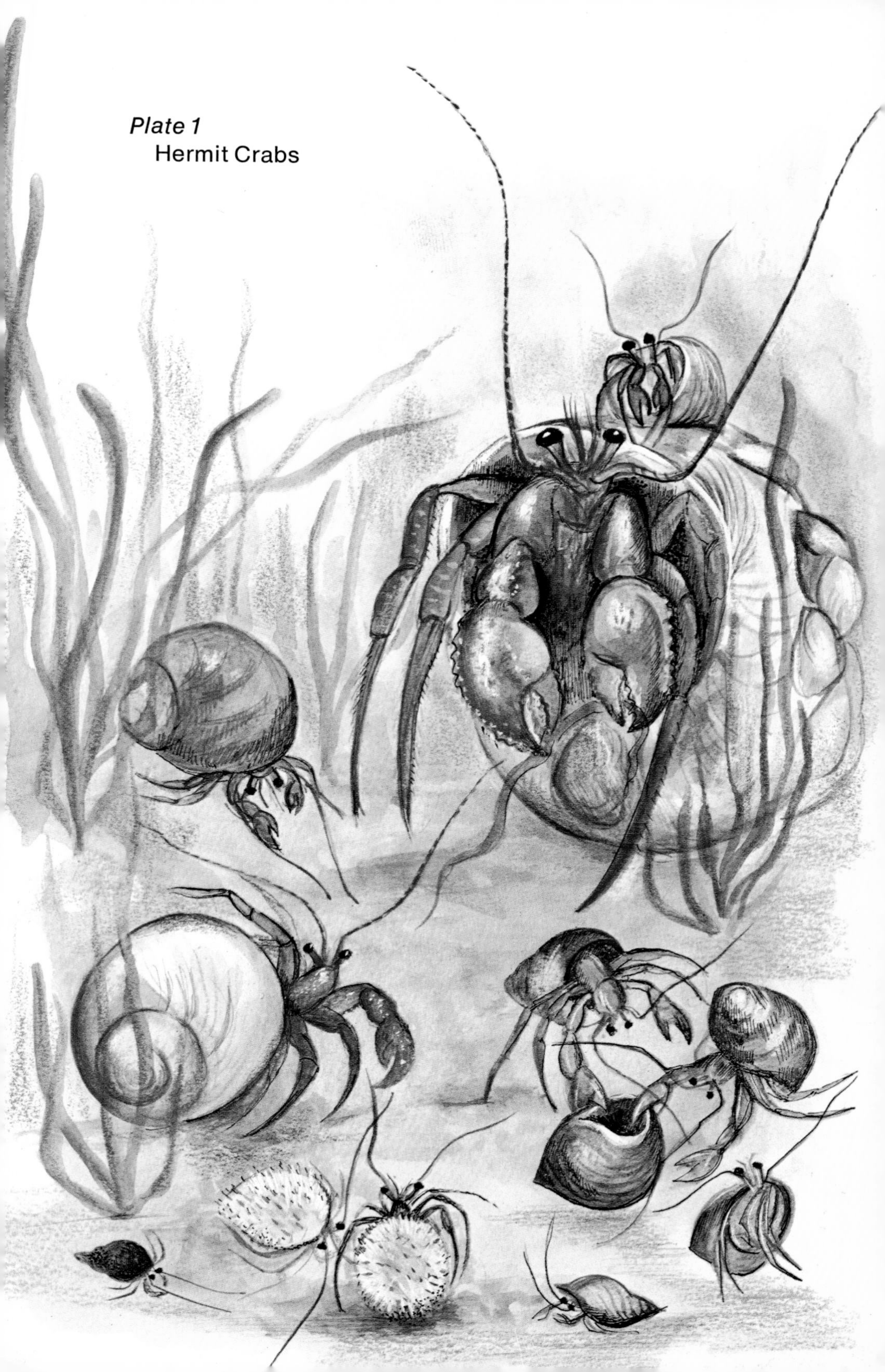
Plate 1
Hermit Crabs

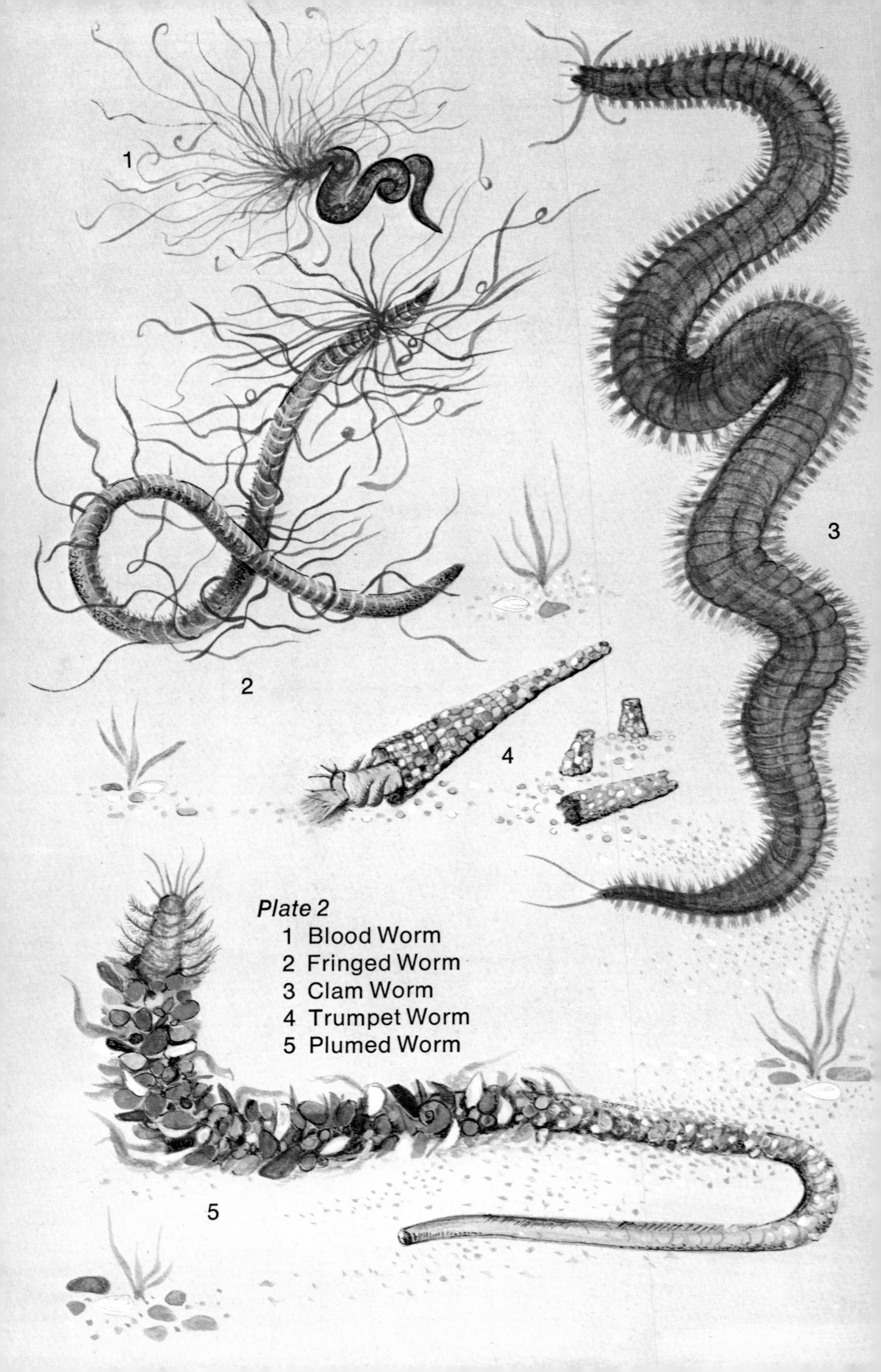

Plate 2

1 Blood Worm
2 Fringed Worm
3 Clam Worm
4 Trumpet Worm
5 Plumed Worm

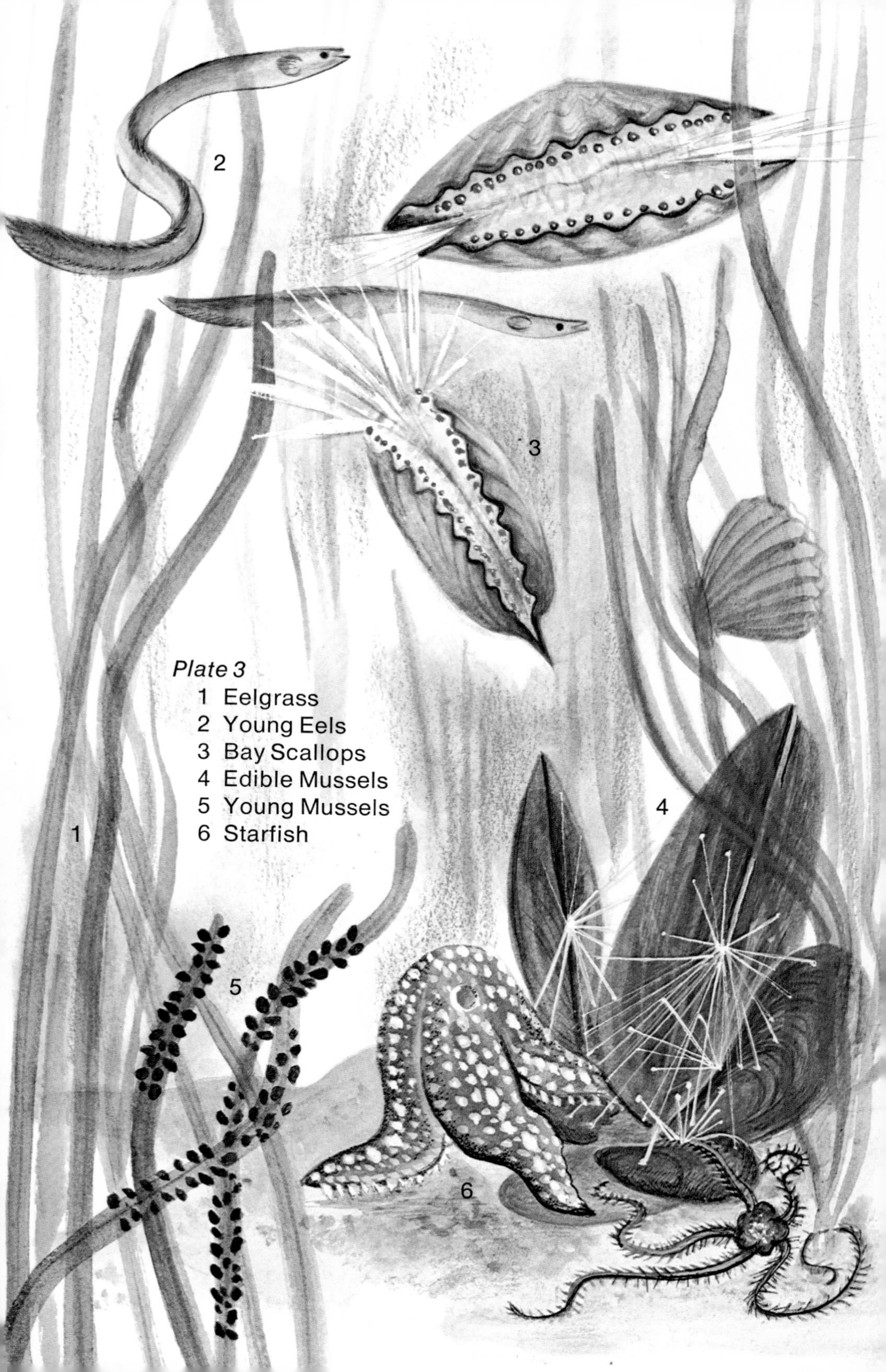

Plate 3
1 Eelgrass
2 Young Eels
3 Bay Scallops
4 Edible Mussels
5 Young Mussels
6 Starfish

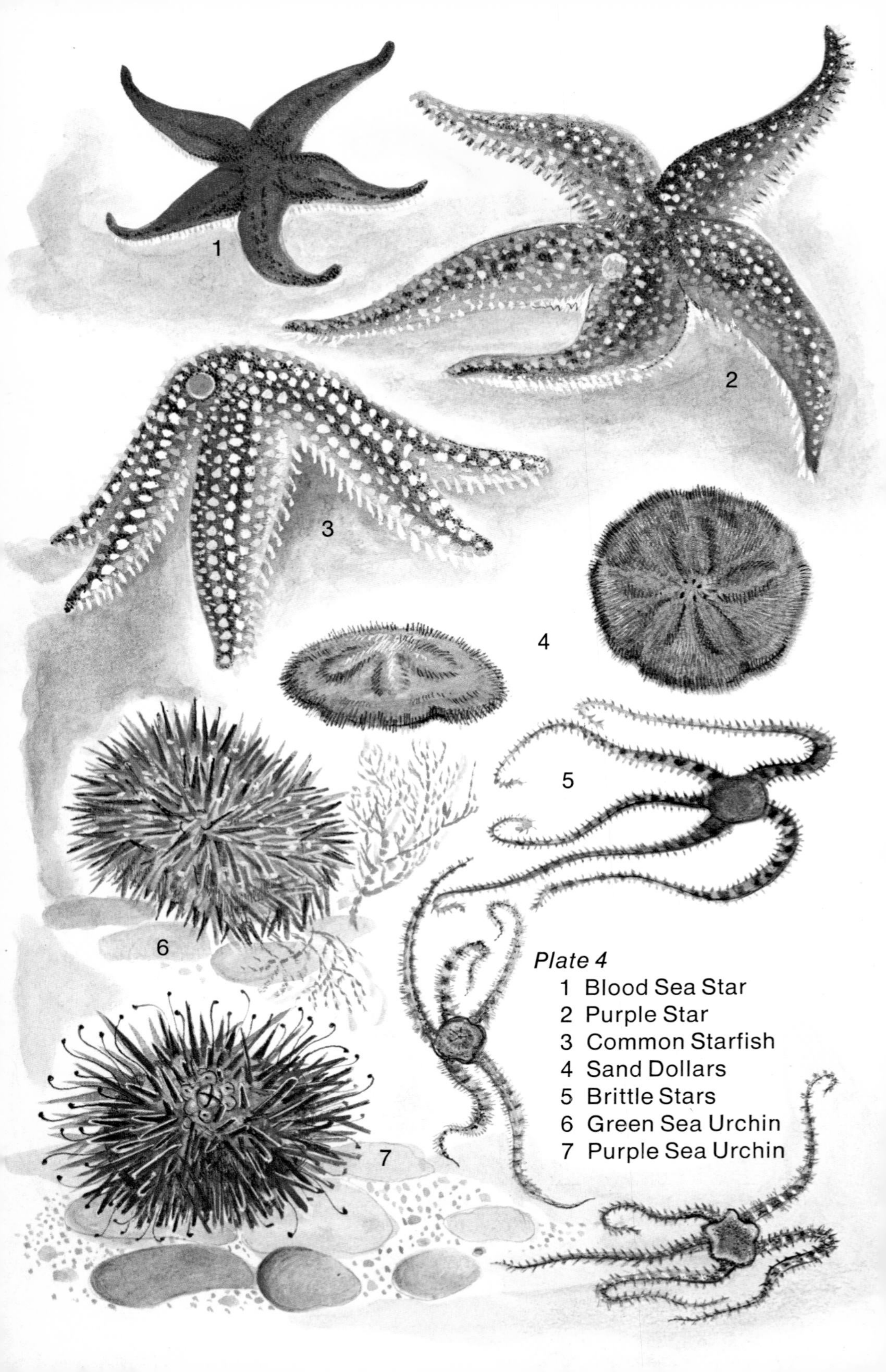

Plate 4

1 Blood Sea Star
2 Purple Star
3 Common Starfish
4 Sand Dollars
5 Brittle Stars
6 Green Sea Urchin
7 Purple Sea Urchin

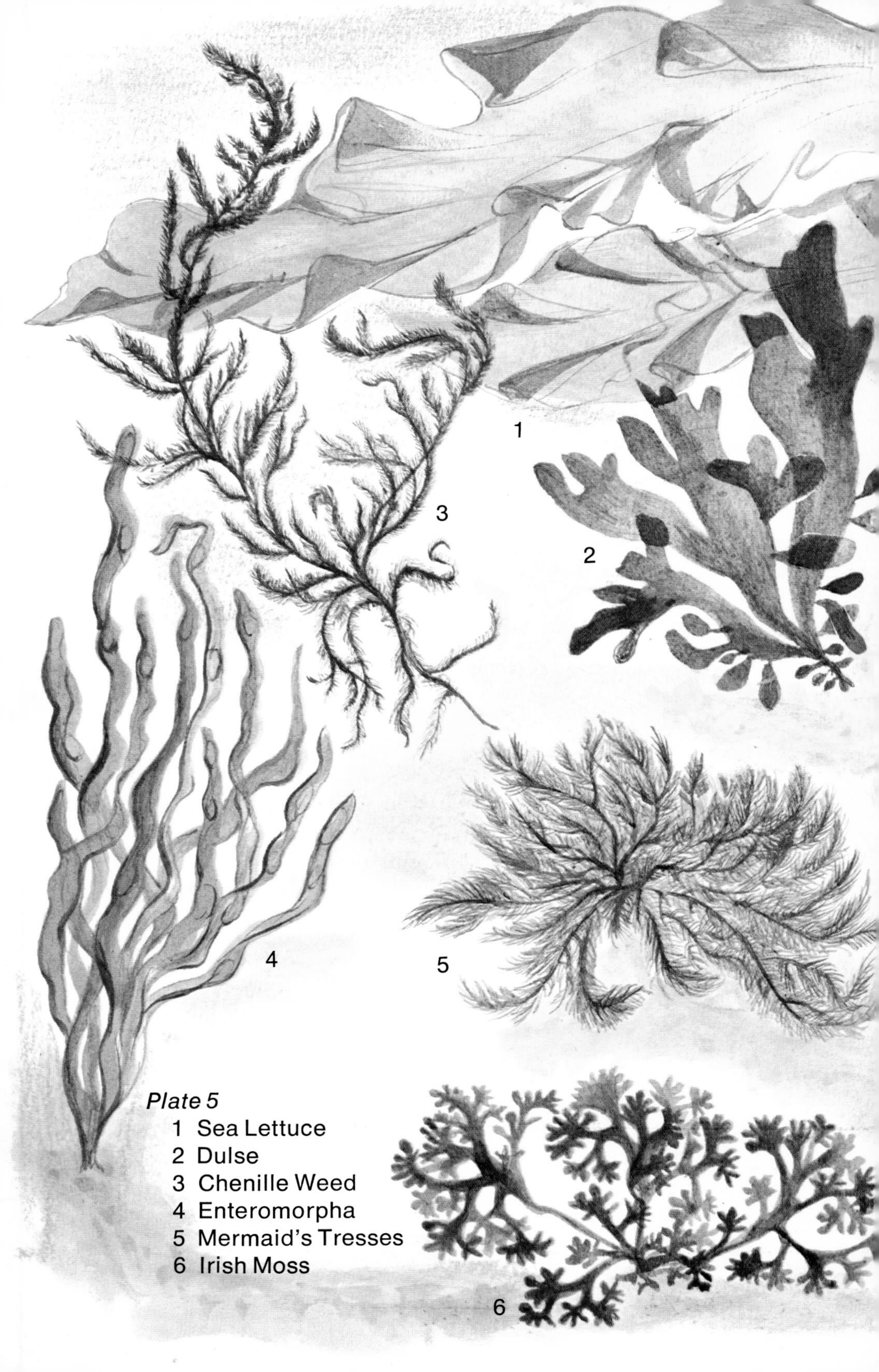

Plate 5
1 Sea Lettuce
2 Dulse
3 Chenille Weed
4 Enteromorpha
5 Mermaid's Tresses
6 Irish Moss

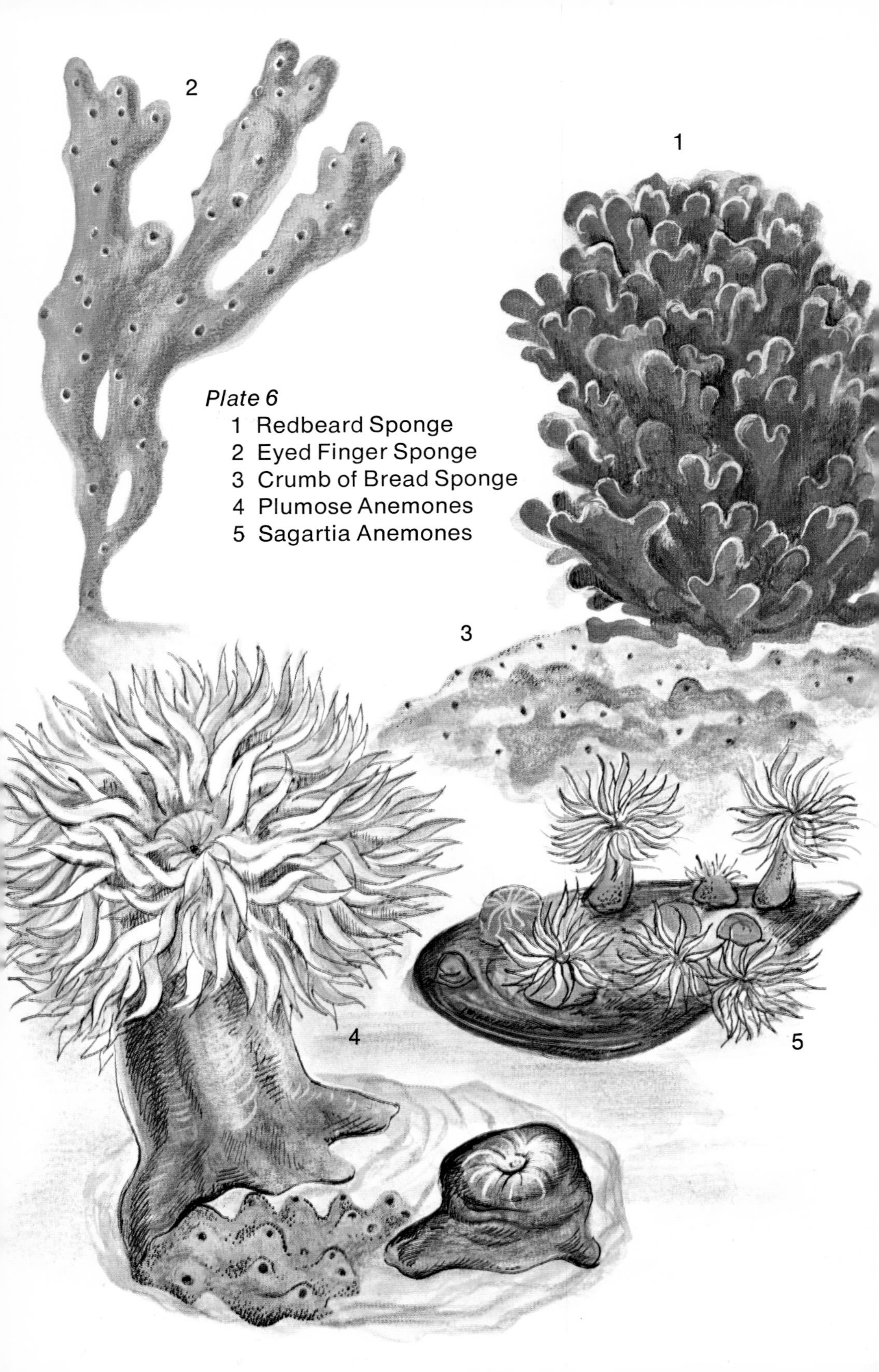

Plate 6

1 Redbeard Sponge
2 Eyed Finger Sponge
3 Crumb of Bread Sponge
4 Plumose Anemones
5 Sagartia Anemones

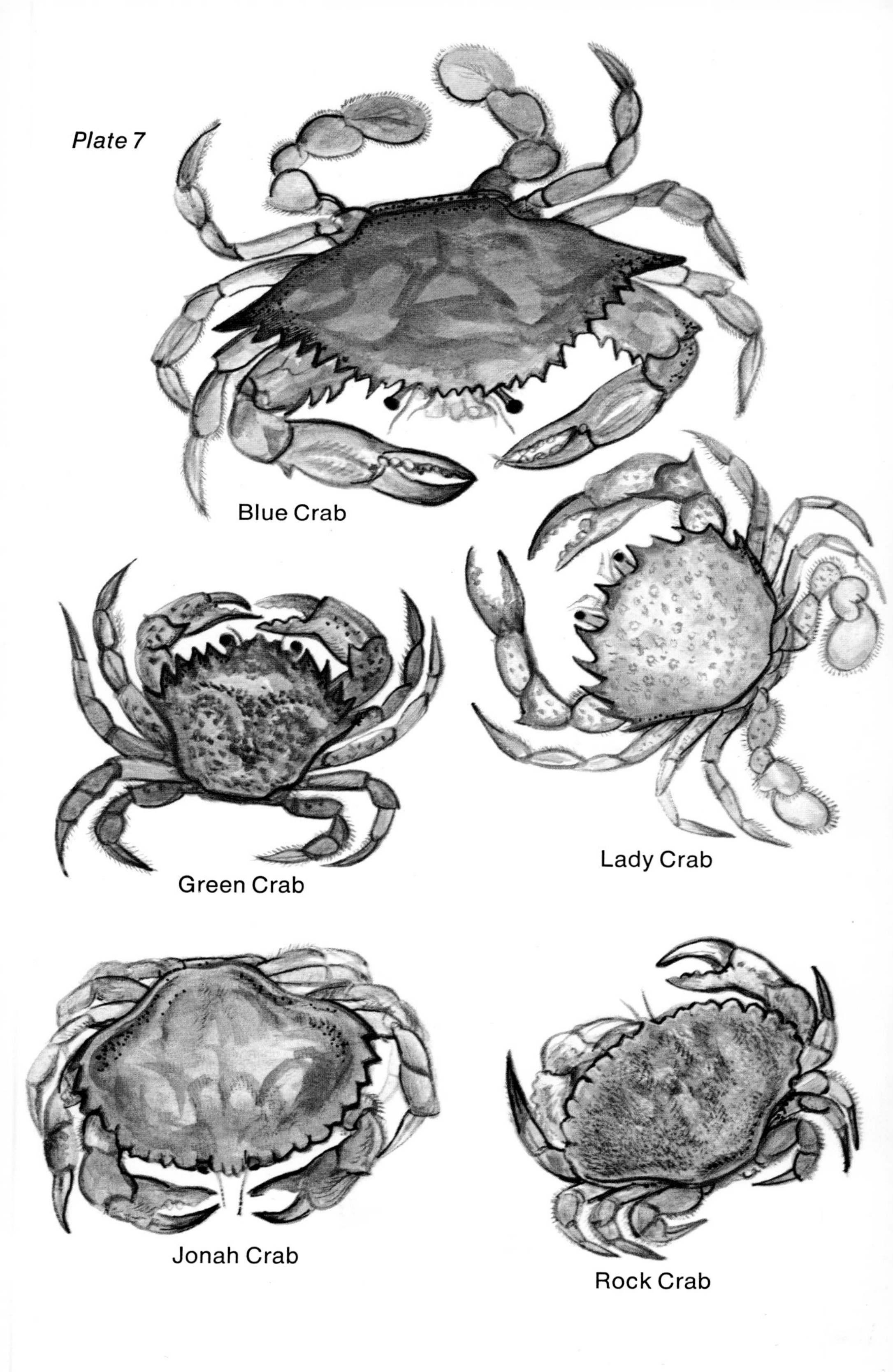
Plate 7
Blue Crab
Lady Crab
Green Crab
Jonah Crab
Rock Crab

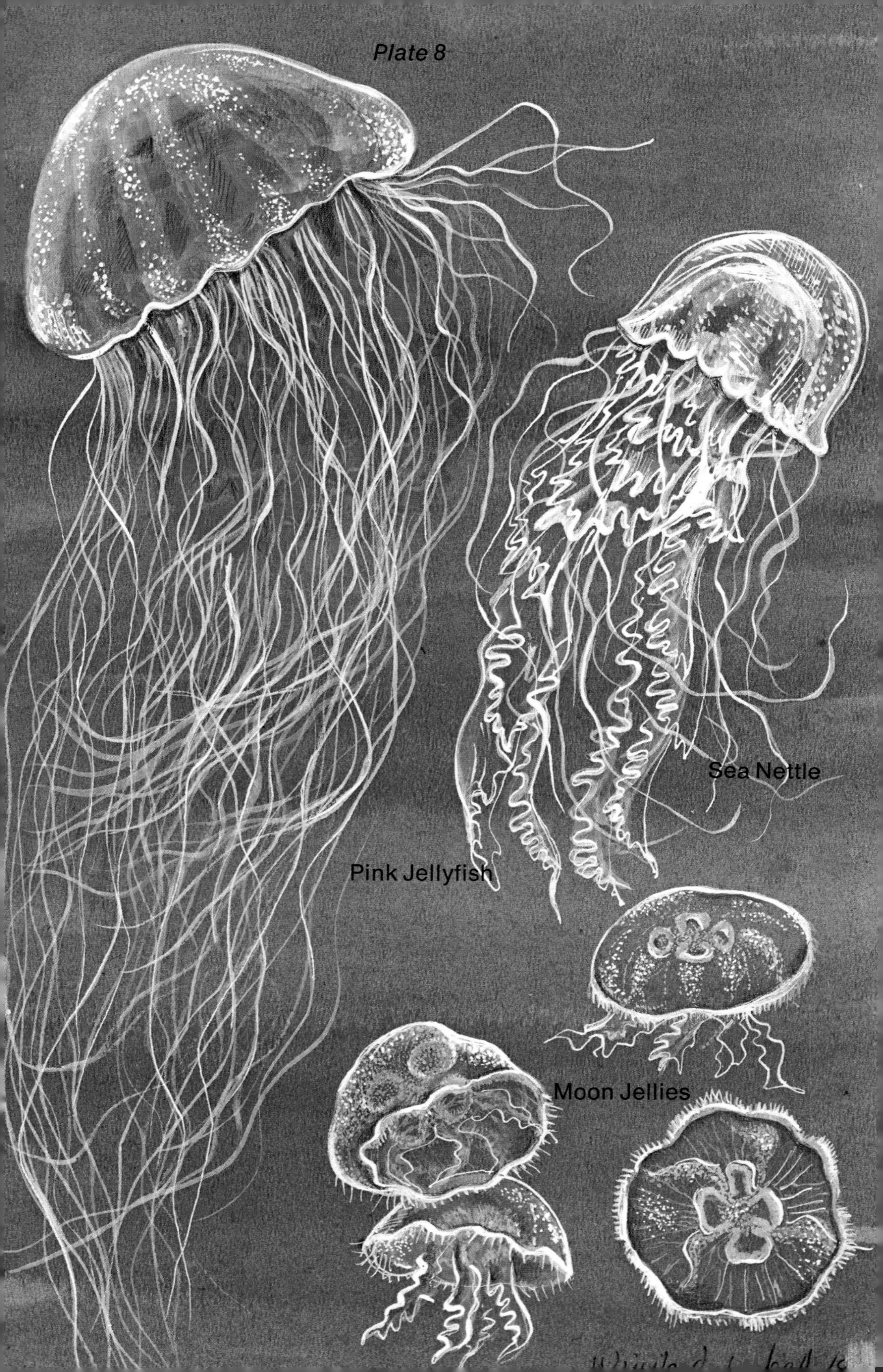
Plate 8
Sea Nettle
Pink Jellyfish
Moon Jellies

1
2
3
4
5
6
Plate 9
1 Herring Gull
2 Laughing Gull
3 Black-backed Gull
4 Common Terns
5 Herring Gull (immature)
6 Black Skimmer

Plate 10
1 Marsh Hawk
2 Canada Goose
3 Black-crowned Night Heron
4 Clapper Rail
5 Black Ducks

Plate 11
1 Sea Lavender
2 Sharp-tailed Sparrow
3 Seaside Gerardia
4 Saltwort
5 Woody Glasswort
6 Fiddler Crabs
1
2
3
4
5
6

Plate 12
Salt-Spray Rose
Seaside Goldenrod
Beach Pea

Plate 13
1 Beach Plum
2 Song Sparrow
3 Bobwhite
4 Bayberry

Plate 14
1 Beach Grass
2 Beach Heather
3 Golden Aster
4 Bearberry
5 Prickly Pear
6 Toad
1
2
3
4
5
6

Plate 15
1 Meadowsweet
2 Buttonbush
3 Pink Azalea
4 Sheep Laurel

Plate 16
1 Barn Swallow
2 Arrowhead
3 Pickerel-weed
4 Water-Lily
5 Pond-Lily
6 Green Frog
7 Damselflies
1
4
3
2
5
7
6

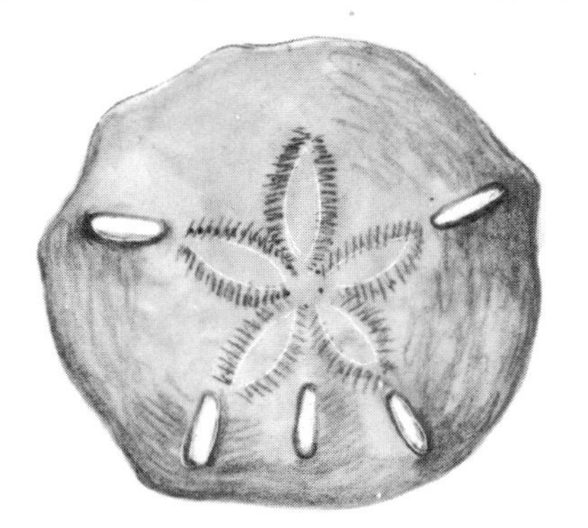

Keyhole Urchin shell

Sand Dollar shell

teeth, protruding from the mouth, are operated by a set of movable jaws. This highly developed chewing apparatus which was first described by Aristotle in the fourth century B.C. is known as Aristotle's lantern.

Two species of sea urchins live in the Outer Lands—the Purple (Plate 4) and the Green Sea Urchin (Plate 4). In spite of their waving bristles and formidable teeth they are not really equipped to do battle. They eat baby mussels and dead fish, but their standard diet is seaweed. In addition to brown algae, they feed on Coralline weeds, getting from them the lime they need for shell-building.

Slow-moving animals, their spines offer them little protection against the gulls, fish and even foxes that hunt them. Hatpin Urchins found in warmer waters are a different proposition, however. Their needle-sharp black spines, which are believed to be poison-coated, can deliver a painful sting.

A Sand Dollar (Plate 4) looks like a flattened sea urchin. Its short spines grow so thickly that the animal seems to be covered with a soft velvet coat. Its mouth, with a simplified version of Aristotle's lantern, is on its underside, while tube feet, projecting from its upper surface, aid in breathing. The perforations for these tube feet form a petal-shaped design on the animal's shell.

The Sand Dollar moves with a wavelike motion of its spines, traveling slowly just under the surface of the sand. It feeds on diatoms and minute plankton creatures which it finds there—and is eaten by cod and flounder. Living at the edge of the flats and in deeper water, Sand Dollars are seldom seen except during the lowest spring tides. Their circular shells wash up on shore after storms.

The common Sand Dollar of the Outer Lands is purple when alive, its spines turning green after exposure to the air. Fishermen used to prepare an indelible ink from these spines by mashing them to a pulp and then adding enough water to make a thin paste. Keyhole Urchins, another form of Sand Dollar, are occasionally found on our shores, but they are more abundant further south. As a Keyhole Urchin glides along on the bottom of the sea, grains of sand sift through the slots on its upper surface. Caught in the animal's spines, the sand helps to conceal it from enemies.

Sea Anemones

Graceful sea anemones cling to the undersurfaces of rocks, seeking shade and moisture behind the seaweed. Although it anchors itself to the rock with a smooth muscular disk, an anemone can move somewhat as a snail does, only more slowly. Its flower-like appearance is deceptive for its "head" consists of circles of waving tentacles. Equipped with stinging cells, the tentacles paralyze small plankton creatures—and sometimes crabs and fish—and carry them to the animal's mouth. Its muscular stalk is little more than a hollow cavity where food is digested and eggs and sperm are produced. Because of their simple body plan, anemones are grouped with the coelenterates, a word that means "hollow gut."

When touched or taken out of water, an anemone contracts. Its stalk shrinks, its tentacles fold in, and a collar covers its mouth. As a squat, almost shapeless

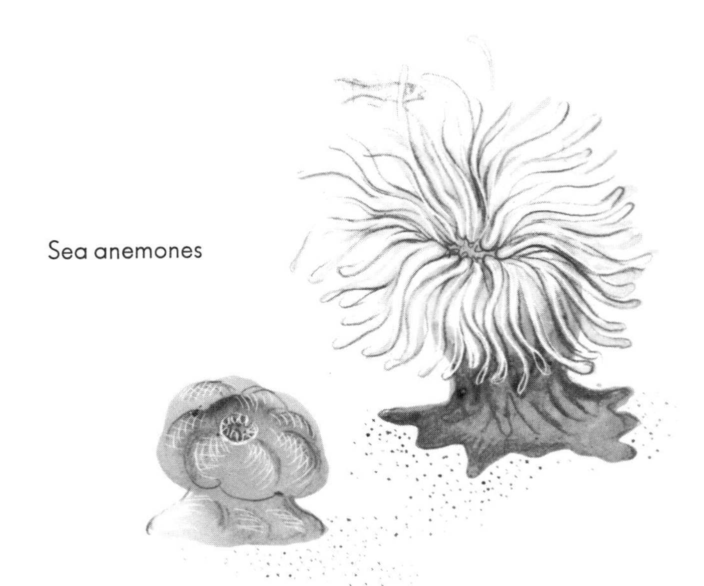

Sea anemones

Star Coral
(greatly enlarged)

blob, it guards against loss of water at low tide. Water is the anemone's primary requirement. It can go for years without food, becoming smaller all the time. If fed regularly, however, it lives indefinitely. The oldest anemones on record were a pair kept at the University of Edinburgh. Fed on liver and hamburger, they lived until they were past eighty. Death came then not from old age, but because of a keeper's carelessness.

Although male and female anemones release clouds of sperm and eggs into the water the animals also increase by fragmentation. As they glide along they leave bits of their footlike disk behind. In time these fragments grow into full-size anemones. They can split in half lengthwise as well, becoming two separate animals in short order.

The frilly Plumose Anemone (Plate 6) lives in northern waters on both sides of the Atlantic and on the Pacific coasts. Growing up to four inches tall (and larger in deep water) it has as many as a thousand short tentacles, which it uses for capturing plankton animals. Several species of Sagartia Anemones (Plate 6) are also found in the Outer Lands. Smaller than the Plumose Anemones their tentacles are long and slender.

Coral

Colonies of Star Coral grow on the undersides of rocks and on stones and shells in shallow water. Closely related to anemones, the small coral animals build cups of lime around themselves. Their stony compartments are a dirty off-white when they wash up on the beach, but they glow with color when the animals are alive. The coral animals usually hide from the light during the day, coming out at night to feed. Less than half an inch high, they thrust out transparent tentacles to catch food. In an aquarium, they will eat fragments of clam and even raw meat.

After hatching from a fertilized egg, a single coral animal starts the colony. Other animals bud from the first one until there are as many as thirty living side by side. Although Star Corals are true corals, their colonies never grow as large as those in tropical waters.

Star Coral skeleton

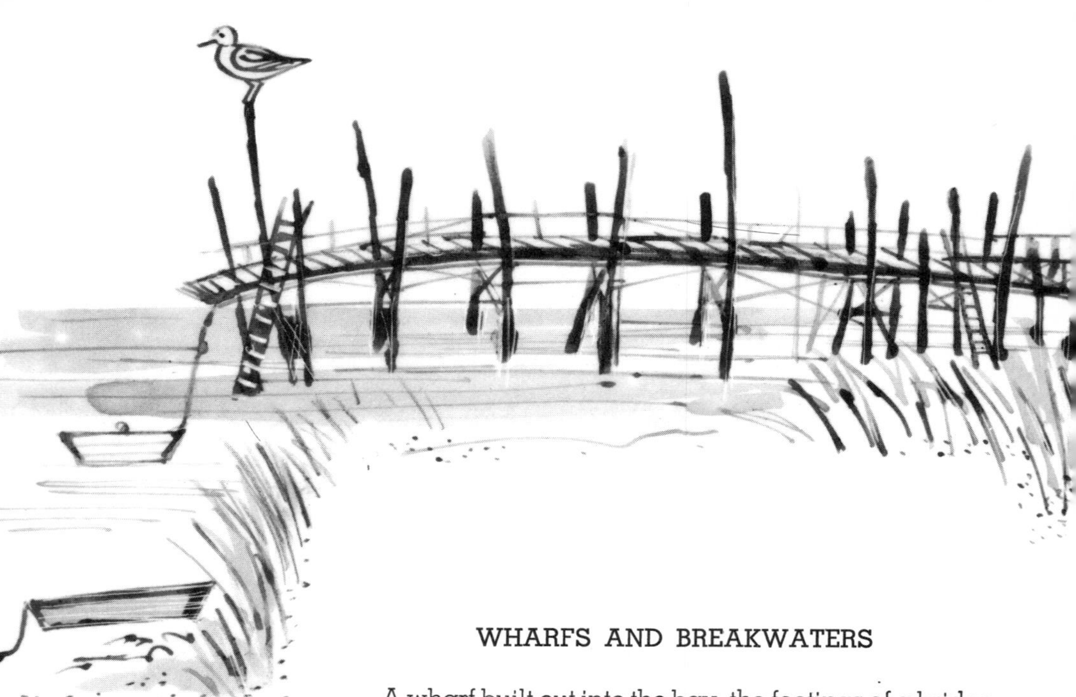

WHARFS AND BREAKWATERS

A wharf built out into the bay, the footings of a bridge across an inlet, a board wedged against the shore—these are invitations to settle down. Decorated with seaweed, studded with barnacles, such wooden structures become tidal cities for small salt-water animals. Mussels and oysters anchor there, attracting a following of starfish and hunting snails, while below the tide line, anemones reach out their tentacles and motionless sponges filter the water.

Sponges

Brightly colored sponges cling to the surface of the pilings. Lacking legs, claws, feelers, mouths, sponges are little more than sieves. Microscopic whips inside their bodies create currents of water. Carrying food particles and oxygen, the water flows into the sponge through thousands of minute pores and out again through larger openings. "To soak up water like a sponge" is more than a convenient phrase, for some of these barely animate creatures are able to "process" hundreds of gallons of water in a day. In shallow tropical seas where the big bath sponges grow, you can actually see the water as it wells out of their openings.

Bath sponges—which are really the skeletons of sponge animals—are made up of an elastic horny substance called spongin. Although most of the sponges of our shores contain spongin, their body walls are also stiffened with minute bits of lime or silica. These scratchy particles make the sponges valueless commercially. Sponges grow irregularly, forming flat crusts on wave-swept beaches and upright branches or cups where the water is calm. Because their shapes and sometimes their colors vary, depending on their locations, sponges can be difficult to identify unless their internal structure is studied with a microscope.

The Redbeard Sponge (Plate 6) is a splash of crimson when it covers a shell on the flats, but it may be six inches high with dozens of stubby branches at the base of a wharf piling in clear water. All sponges produce eggs and sperm, the fertilized egg sailing away from its parents on an outgoing stream of water. They also reproduce by budding, spreading out in all directions over the surface of the piling. Redbeards, in particular, are noted for their ability to regenerate. When a piece of the sponge is squeezed through a fine mesh cloth, the separate cells, visible as bright red dots, settle on the bottom of a dish of sea water. In a few minutes they begin to creep about. Bumping into each other, they clump together to form small balls. Given time and favorable conditions, the balls grow into new sponges.

Although this can be done at home, it is difficult to

Clathria Sponge

keep a sponge alive. If water temperature and other conditions are not just right, the animal dies quickly, its color fading to a dull brown.

The Eyed Finger Sponge (Plate 6), also called Deadman's Fingers, is often washed up on the beach. The conspicuous "eyes" are the exit holes for water and wastes. Orange when alive, the sponge's fingers grow from a single stalk. This is one way of distinguishing it from the Clathria Sponge whose fingers spring up separately from a flattened base. Mermaids' Gloves, another look-alike, has pale yellow fingers which are sometimes a foot long.

The Crumb of Bread Sponge (Plate 6) forms fleshy yellow-green patches only a fraction of an inch high. The surface of the sponge is dotted with miniature craters through which water oozes out. Softer than most sponges, it crumbles easily. No animal is likely to eat a Crumb of Bread, however, for it has a strong, disagreeable odor.

Mermaids' Gloves

Sea Squirts

Sea Squirts

Sea Squirts cluster together on pilings or hang from the undersides of floats. A typical Sea Squirt looks like a small plastic bottle with two spouts. One spout is the animal's mouth, the other the opening through which water leaves its body. When disturbed, the animal contracts, squirting water out of both openings. The rounded Sea Grapes grow singly, but some squirts are colonial animals, budding to form good-sized colonies. Irregular masses of Sea Pork may be eight inches long. Smooth and shiny, they resemble chunks of raw salt pork.

Although Sea Squirts are as stationary as sponges, they are far higher on the evolutionary scale. Their tough outer covering consists largely of cellulose, the same material that cell walls of plants are made of. This outer coat is known as a tunic, so that Sea Squirts are often called tunicates. Underneath the tunic beats a remarkable heart. After pumping blood in one direction, the heart pauses, goes into reverse, and pumps it the other way. Sea Squirts also have a full complement of digestive and reproductive organs as well as simple brains.

These odd animals are of special interest to scientists because of their young. Lively swimmers, about an eighth of an inch long, they look like frog tadpoles with exceptionally long tails. Inside the tail is a primi-

Sea Grapes

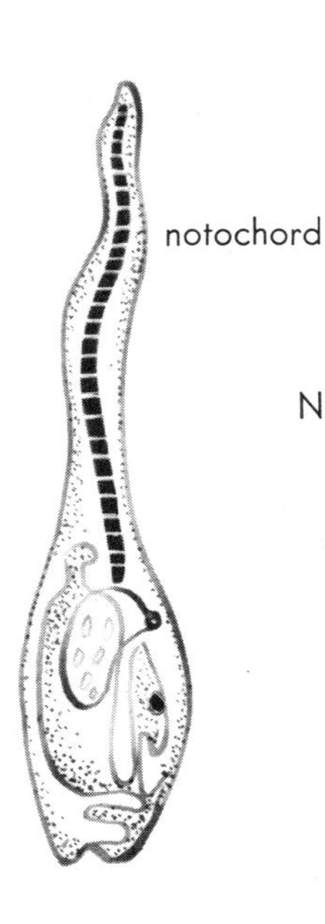

Newly hatched Sea Squirt

tive but unmistakable backbone, called a notochord. The tail and notochord disappear after the young squirt fastens to a wharf piling, but they are present long enough to place these animals as chordates, a group that also includes man. The discovery of the Sea Squirt's backbone, made only a few years after Darwin's theory of evolution was published, was an important link in the chain of evidence that proved the correctness of the theory.

Shipworms

Shipworms live inside of wharf pilings, digging tunnels in the water-soaked wood. They are not worms, but two-shelled mollusks, related to Angel Wings. After a week or two in the sea, a young Shipworm attaches itself to a wooden surface with a single byssus thread. Its shells change shape, becoming curving blades with rows of fine teeth along their edges. The animal uses these ridged shells to carve out a cylindrical tunnel up to two feet long. The longer the tunnel, the longer the Shipworm, for its siphons remain at the tunnel entrance to take in food and water and discharge wastes. The Shipworm also gets nourishment from the wood it scrapes away. The "sawdust" is swallowed and at least some of it is digested as it passes through the animal's long gut.

The tiny entrance holes on the surface of a timber give little hint of the destruction underneath. Shipworms tunnel rapidly, boring inch-long holes in little

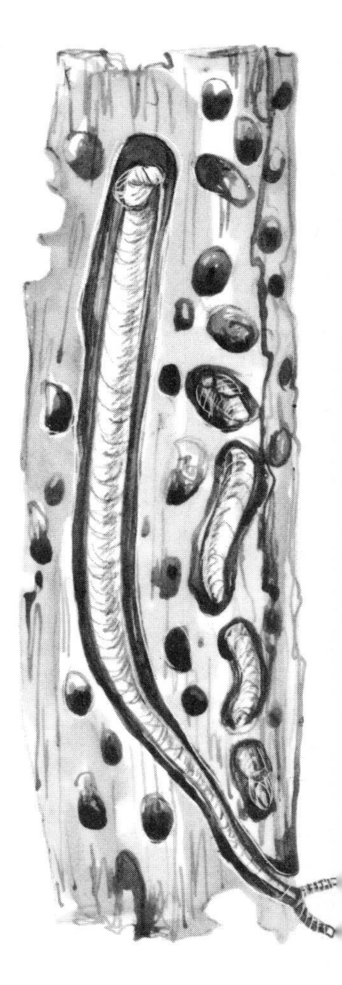

more than a day. In six months they may honeycomb a piling so completely that the wood disintegrates. The Shipworms die once their burrows are exposed but by that time they have produced millions of offspring to carry on their work.

Before the days of steel hulls, Shipworms traveled from port to port by boat. They riddled the planking of Greek galleys, destroyed the caravels of early explorers, and almost flooded Holland by boring through its dikes. When they first appeared in San Francisco during World War I, wharves collapsed without warning, tumbling warehouses and freight trains into the bay. Treating pilings with creosote or anti-fouling paints protects the wood for a time, but the centuries-old battle with these "termites of the sea" is far from won. The Navy estimates that the damage they do to boats, docks and bridges in the United States amounts to more than $50 million a year.

Shipworms in wood

Green Crab

SHALLOW WATER

The shallow seas in which life on our planet originated are often described as an "organic soup." In contrast to those ancient waterways, our bays and sounds are like a rich chowder. Infinite numbers of young clams, snails and worms join the plankton wanderers for a day or a week, before settling down on the bottom. But the waters also teem with larger creatures —scuttling crabs, pulsing jellyfish, schools of streamlined squid. Some of these creatures come ashore. Others must be pursued with dip nets or observed in the water through goggles or face masks.

Crabs

Basically, a crab is a short-tailed lobster. It has similar jointed legs and heavy one-piece shell, but where the lobster's abdomen is long and flexible, the crab's is only a short flap which folds under its body. An active animal, it walks along the bottom of the bay or uses its legs for swimming. Its diet is varied, ranging from seaweed to live fish.

A crab molts as a lobster does. Its shell splits along the rear edge and the animal slowly backs out. During

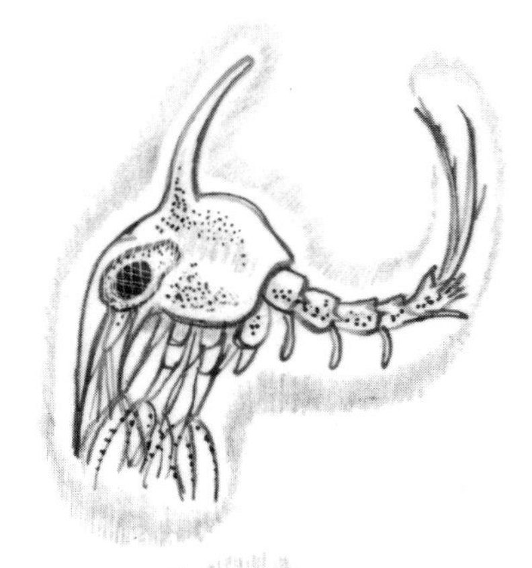

Young Rock Crabs

the hours afterward, before its new shell has had a chance to harden, it is a soft-shelled crab. Every species of crab is a soft-shell many times during its life, but only the Blue Crab (Plate 7) is sold when its shell is papery. A commercial fisherman can tell by a fine line around the rim of the shell when molting is about to take place. He keeps these "peelers" in special pens so that he can market them as soon as they have completed their molts.

The molt is a signal to the male Blue Crab as well as to the human predator with net and frying pan. Two or three days before a female molts, a male begins to court her. He dances in front of her with outstretched claws and then carries her around, holding her underneath his own body with his claws and front legs. While she molts, he hovers over her. They mate soon afterward. During the next two days while her shell is hardening, he continues to carry her underneath him. After that, she is on her own.

A female Blue Crab mates only once, obtaining

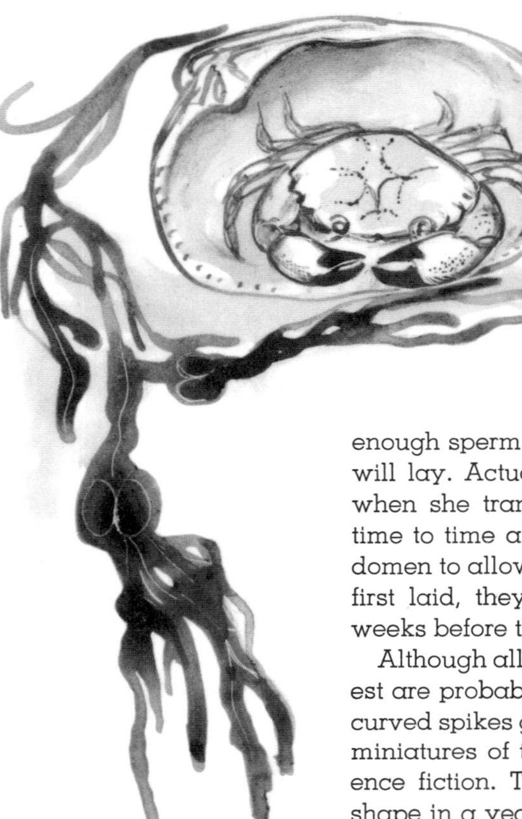

Mud Crab

enough sperm then to fertilize the millions of eggs she will lay. Actual egg-laying takes place months later when she transfers the eggs to her abdomen. From time to time as she swims, she lowers the folded abdomen to allow water to reach the eggs. Orange when first laid, they become dark brown during the two weeks before they hatch.

Although all marine young are odd-looking, the oddest are probably the crabs. Big-eyed, long-tailed, with curved spikes growing from their heads, they resemble miniatures of the men from Mars in the pages of science fiction. The young crabs reach adult size and shape in a year or a little longer. Some live to be two years old, but only a few reach the ripe old age of three.

Blue Crabs belong to the family of Swimming Crabs. Their flattened hind legs, shaped like paddles, permit them to swim rapidly—sideways, backward, and sometimes forward. The speckled Lady Crab (Plate 7) is also a swimmer who skirts the shore to pick up clam meat or small fish. With a face mask you can easily watch it feed. An aggressive little animal, it will snatch a broken clam from the claws of a Hermit or chase away a group of nibbling shrimp. Any movement on your part will bring the undersea drama to an end as the crab hastily backs into the sand, buried except for its stalked eyes and waving claws. These menacing pincers, raised and ready to attack, give the animal the reputation of being "crabby."

Green Crabs (Plate 7) are also members of the Swimming Crab family, although their hind legs are pointed rather than paddle-shaped. They burrow into the sand to find food, eating worms and young Soft-shell Clams. Crushing the clam shells with their claws, they can eat fifteen in a day. Formerly Green Crabs were never found north of Cape Cod, but as coastal waters have grown warmer, they have moved all the way to Canada. In Maine they have become the major pest on clam beds. Considered good eating in Europe, they are caught here only for bait.

The Cancer Crabs (*cancer* is the Latin word for crab) are usually larger than Swimming Crabs, with more nearly oval shells. Walkers rather than swimmers, they are found in rocky places. They lie half-buried under stones or gravel at low tide. The Common Rock Crab (Plate 7) lives in sheltered bays while the brick-red Jonah Crab (Plate 7) is found on outer shores where the surf is heavy.

Spider Crab

Mud Crabs crawl across the flats in sandy and muddy areas. Between tides they take shelter in empty clam and oyster shells. Olive brown, often with black-tipped claws, they are not much larger than Hermits. Despite their size, they are more pugnacious than their neighbors, the Spider Crabs. These long-legged, slow-moving creatures which look like huge spiders are easy targets for gulls and fish. Their lives depend on good camouflage. The Common Spider Crab of the Outer Lands decorates itself with bits of seaweed which it picks up with its claws and impales on a row of hooked spines on its back. The seaweed—and even sponges and barnacles—sometimes grows on the crab's rough shell.

Horseshoe Crabs

No one is likely to write a poem in praise of the Horseshoe Crab. A mysterious-looking creature in brown armor, it could seem frightening on first encounter. Once you know something of its life, however, it takes on an awkward charm of its own. In any event, its ancient lineage must command respect.

The Horseshoe Crab—which is not a crab at all—has

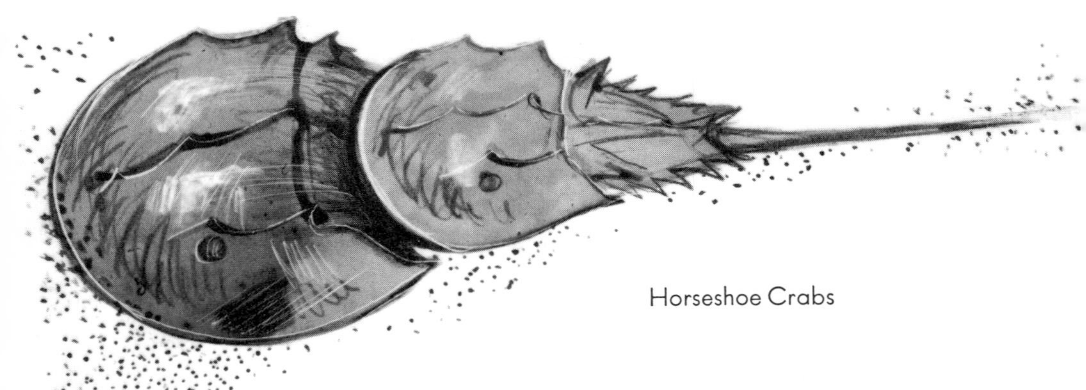

Horseshoe Crabs

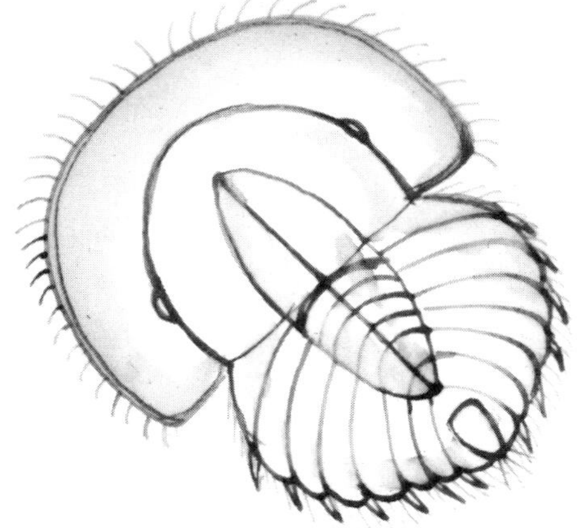

Newborn Horseshoe Crab
(greatly enlarged)

outlived the ichthyosaur, the pterosaur and the stegocephalian. These were its companions when it first plodded up the beach some 200 million years ago. As mountains rose and seas receded, as climate turned about, its contemporaries of later geologic ages also became extinct. We know of them only from the fossil records, but Horseshoe Crabs live on. From the Gulf of Maine to the Gulf of Mexico, they crawl along the bottoms of bays much as they did in primeval times. Our Horseshoe Crab is found only on the western shores of the Atlantic Ocean, although related species live in Asia.

In May and June when spring tides flood the beaches, Horseshoe Crabs move ashore. They come two by two, a female leading while a smaller male clings to her armored shell. At the high-tide line, she digs a shallow nest in the sand. As she lays her eggs, the male fertilizes them. With the ebbing tide, they lumber back to sea.

Sand drifts over the eggs. The sun warms them. For two weeks the developing embryos remain high and dry on shore. When the next spring tides come, waves uncover the nest and carry the baby Horseshoes to the water. Pale tailless creatures, they burrow into the sand of the flats. Molting from time to time, they grow very slowly. As yearlings they are only an inch wide; at three, their shield-shaped shells are scarcely three inches across. After they molt, their cast-off shells drift to shore. If you examine one closely you will see that

Cast-off
Horseshoe Crab shells

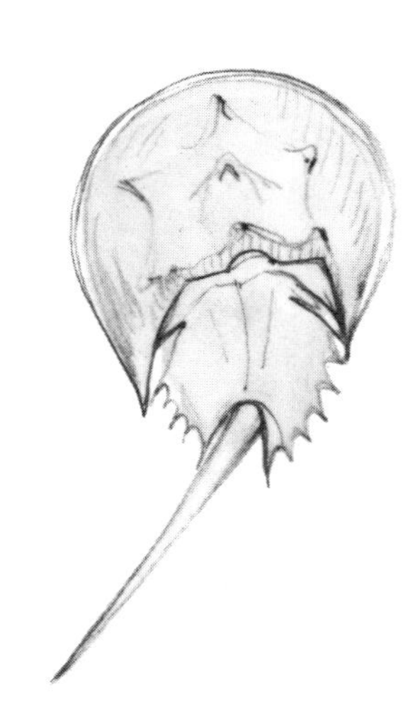

it has a slit all along the front edge. Unlike lobsters and crabs, the Horseshoes crawl out of their shell head-first.

During their growing years, as their armor hardens and turns brown, they leave the flats for deeper waters. Sometimes they travel from bay to bay in search of food. Although they are not long-distance travelers, one tagged Horseshoe Crab was caught twenty-one miles away from the beach where it was first seen.

The animals mature between the ages of nine and twelve and may live to be nineteen. After they are full-grown, they start their yearly trips to shore. As they bob along through the shallow water, they resemble walking zoos. Flatworms cling to their undersides, while Boat Snails, barnacles and Tube Worms grow on their heavy shells.

Descendants of the giant water scorpions who dominated the seas during the Paleozoic era, Horseshoe Crabs are related to modern land scorpions and spiders. Their hinged two-part shells end in a pointed tail —a tail that helps the animal to right itself if it is turned over, but is not used for defense or attack. Large compound eyes and a pair of smaller eyes are located far back on the domed shell. Looking upward, the compound eyes continue to see even when the animal is half-buried in the sand.

Scientists have found that these eyes are part of a surprisingly complex nervous system. If a Horseshoe Crab is put down on the beach, it will immediately head toward the bay, even if obstacles prevent it from

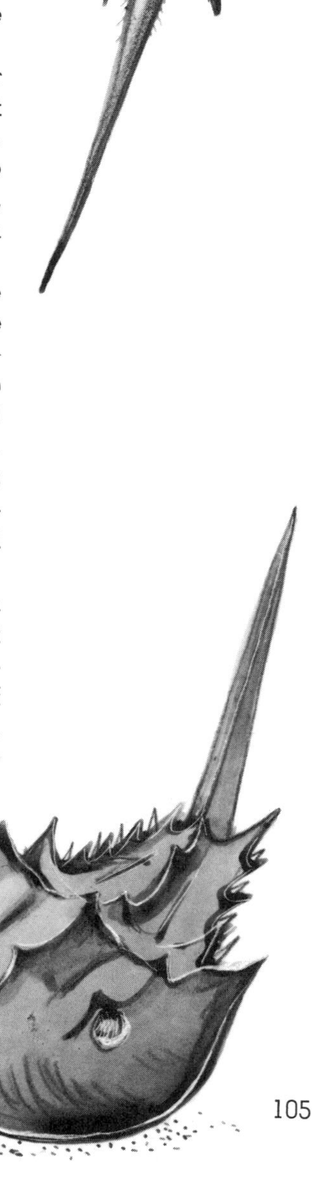

seeing the water. Only on a cloudy day or when its eyes have been covered will it head the wrong way. Apparently it takes its bearings, as birds do, from the sun and the patterns of light in the sky.

On its underside, a Horseshoe Crab has a small pair of pincers and ten legs. The first four pairs of legs lift the animal up while the fifth pair shoves it forward, giving it an odd, jerky gait. The legs do a double job because the Horseshoe, which has no teeth or jaws, crushes its food with short spines at the base of each leg—as if it were chewing with its shoulders.

Equally unusual are the flapping gill books on the animal's abdomen, so-called because they are made up of hundreds of papery leaves. These are the Horseshoe's breathing apparatus, but they also help it to swim. During the day it uses the curved front of its shell to plow through the sand. At night it goes swimming. Taking off from the bottom of the bay, it slowly flips over on its back. Then leg and gill books move in unison, like the oars of a varsity crew, and the upturned animal glides gracefully through the water.

Horseshoe Crabs have been around for 200 million years, but they may not continue to be. Unlike most shore animals, they mature slowly and lay only limited quantities of eggs. Fish and birds eat perhaps half their eggs and quantities of young Horseshoes as well. When the animals grow older, however, their armor

Moon Jellies developing

protects them from all enemies except man. Along stretches of the New Jersey shore millions of Horseshoes were trapped annually, to be used as chicken feed and fertilizer, until so few remained that trapping was no longer profitable. A similar fate may await them on the Outer Lands where shell-fishermen dislike them because they eat Soft-shell Clams. Although it is questionable whether they kill as many clams as careless clam-diggers do, they have been declared outlaws in some areas. At least one Cape Cod town encourages their destruction by paying a two-cent bounty for their carcasses—which seems an extraordinarily low price to set on our most ancient senior citizens.

Jellyfish

Although jellyfish are creatures of the open sea, great numbers of them drift to shore during the summer. Relatives of sea anemones and hydroids, their simple bodies consist of two layers of cells enclosing a jellylike substance. The blue-white Moon Jelly is bell-shaped, with a fringe of stinging tentacles and four dangling mouth parts. It swims by a series of rhythmical pulsations which are almost like heartbeats. When its bell contracts, forcing out water, the jellyfish moves upward. As the bell expands again, the animal slowly sinks down. On its downward trip, with the bell spread out like a parachute, it snares the plankton animals on which it feeds. Simple sense organs on the margin of the bell control its movements.

Stalked Jellyfish
(enlarged)

A Moon Jelly (Plate 8) can be recognized by the cloverleaf pattern on its upper surface. These are its reproductive organs—white if the animal is female, pink if a male. After the eggs are fertilized, the developing young cling to the folds of the female's mouth parts. Shaken loose late in summer, they attach to rocks or wharf pilings. Small plantlike creatures, no more than a quarter of an inch long, they hang there during the winter, using slender tentacles to catch food. In the spring their bodies change form until they look like stacks of saucers. One by one the "saucers" break away from each other and become tiny Moon Jellies. Growing quickly, they reach full size—up to ten inches in diameter—by mid-summer, and the cycle starts all over again. This method of reproduction, known as an alternation of generations, is typical of jellyfish and many of their hydroid relatives.

Moon Jellies and the small stalked jellyfish that fasten to blades of Eelgrass and seaweed can be handled safely, but it is wise to head away from other large jellyfish that swim offshore. A Sea Nettle (Plate 8) has long golden tentacles and clusters of stinging cells scattered over the surface of its bell. These stingers discharge poison into anything they touch. Acting something like curare, the poison paralyzes and then kills small crustaceans and fish and raises painful welts on the arms of human swimmers.

The Pink Jellyfish (Plate 8), also called the Red or Arctic Jellyfish, is larger and more dangerous. In cold northern waters it can be eight feet in diameter with

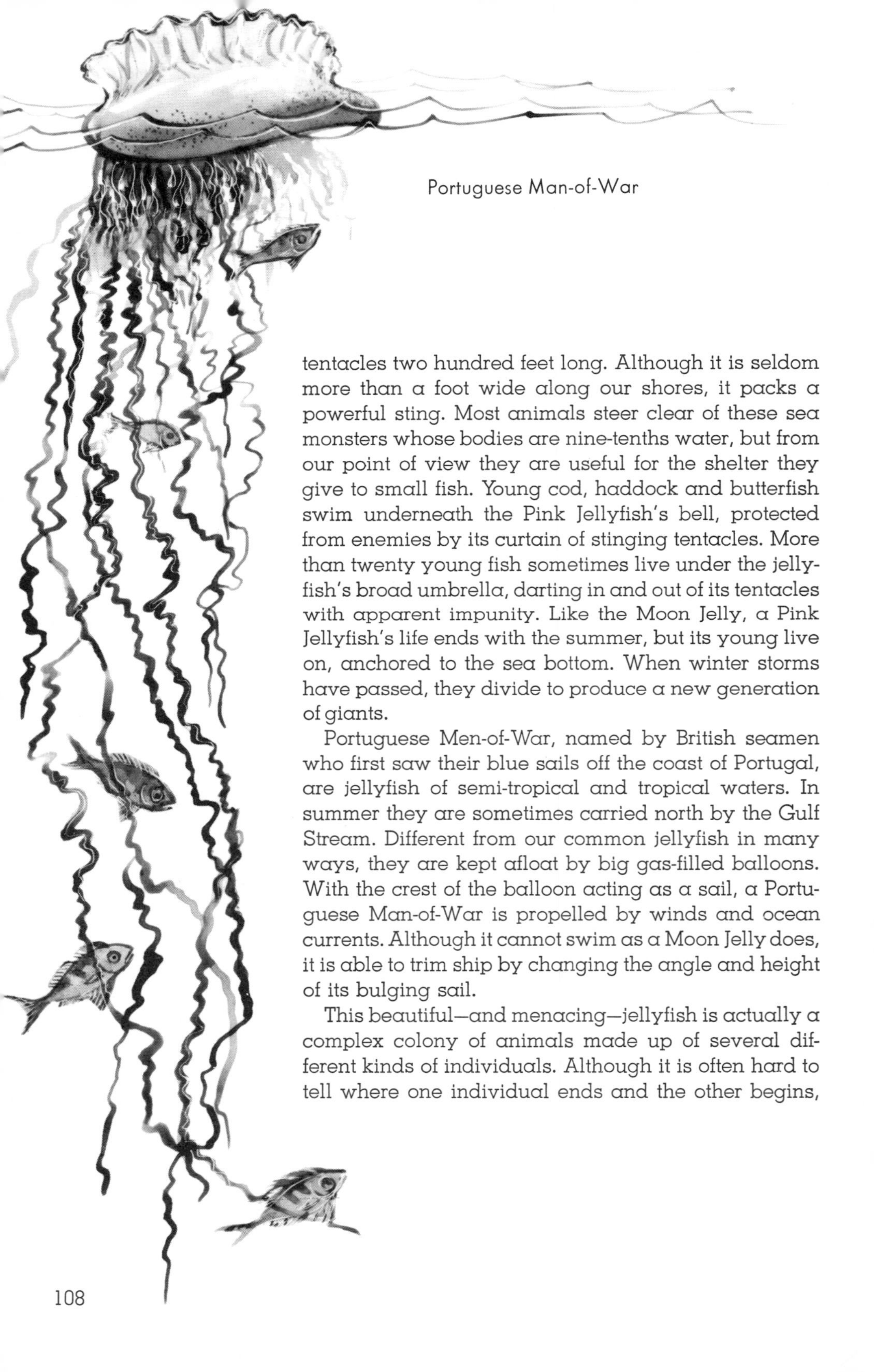

Portuguese Man-of-War

tentacles two hundred feet long. Although it is seldom more than a foot wide along our shores, it packs a powerful sting. Most animals steer clear of these sea monsters whose bodies are nine-tenths water, but from our point of view they are useful for the shelter they give to small fish. Young cod, haddock and butterfish swim underneath the Pink Jellyfish's bell, protected from enemies by its curtain of stinging tentacles. More than twenty young fish sometimes live under the jellyfish's broad umbrella, darting in and out of its tentacles with apparent impunity. Like the Moon Jelly, a Pink Jellyfish's life ends with the summer, but its young live on, anchored to the sea bottom. When winter storms have passed, they divide to produce a new generation of giants.

Portuguese Men-of-War, named by British seamen who first saw their blue sails off the coast of Portugal, are jellyfish of semi-tropical and tropical waters. In summer they are sometimes carried north by the Gulf Stream. Different from our common jellyfish in many ways, they are kept afloat by big gas-filled balloons. With the crest of the balloon acting as a sail, a Portuguese Man-of-War is propelled by winds and ocean currents. Although it cannot swim as a Moon Jelly does, it is able to trim ship by changing the angle and height of its bulging sail.

This beautiful—and menacing—jellyfish is actually a complex colony of animals made up of several different kinds of individuals. Although it is often hard to tell where one individual ends and the other begins, some are responsible for capturing food, others for eating, and still others for defense and reproduction. Tentacles up to fifty feet long trail from the underside of the brilliant red and blue float. They wrap around a fish, paralyzing it with their stinging cells, and then carry it up to the animal's many mouths.

The Portuguese Man-of-War's stinging cells contain the most powerful poison known in marine animals. A single sting is painful and a swimmer who becomes entangled in its trailing tentacles can be seriously hurt. Even the tentacles of a dead Man-of-War on the beach retain their poison for a long time.

In spite of their deadliness, these floating fortresses shelter one species of small fish, the *Nomeus*, which is never found away from the company of its protector. Some scientists believe that *Nomeus* serves as live bait, luring larger fish to the Man-of-War's tentacles.

Squid

From time to time schools of squid enter shallow water as they chase young mackerel or flee from big striped bass. There is something uncanny about the way these animals swim, with their arms extended and their wise eyes focused on their prey. Squid are as mobile as the fish on which they feed, but are totally unrelated to them. Instead they are mollusks whose ties are with the oyster and snail.

In place of the oyster's heavy shell, a squid has only a narrow horny plate buried within its mantle. The

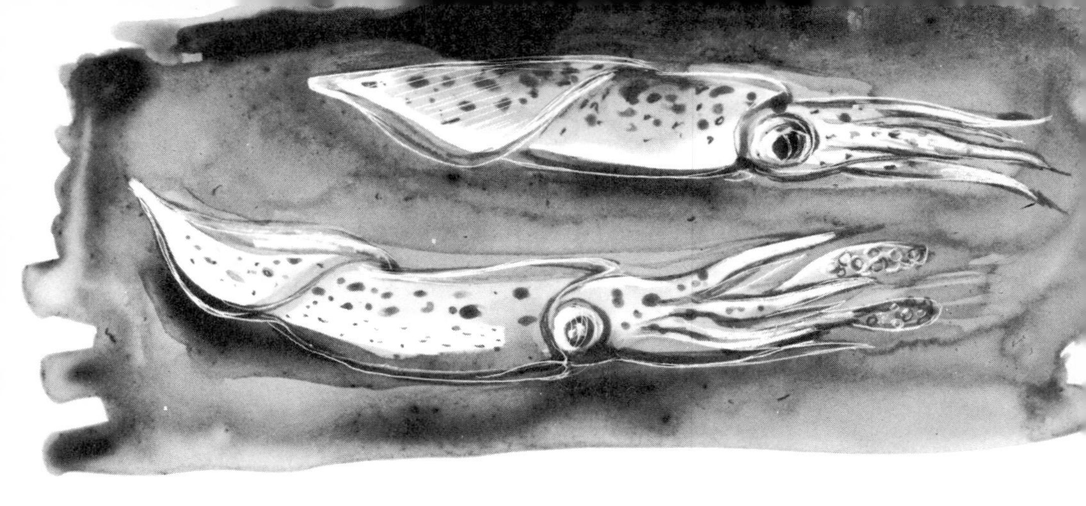

Squid

mantle, thick and muscular, serves as the animal's outer covering. In place of a snail's foot, a squid has ten arms extending from its head—which is why squid are classified as cephalopods, meaning "head-footed." The two longer arms are tentacles, used for capturing fish. The shorter arms hold the food as the squid eats. Although it has a rasped tongue like a snail's, it cuts up its meat with a sharp parrot-like beak. In addition to covering the soft parts of its body, the squid's mantle enables it to swim by jet propulsion as a scallop does, only in a far more sophisticated way. Taking in water at the edge of its mantle, the squid forces it out again through a siphon next to its head. The flexible tip of the siphon can be pointed in different directions, enabling the animal to dart backward and forward and to dip, bank and make U-turns.

The speediest of all invertebrate swimmers, squid also have well-developed brains and nervous systems. Their intelligent-looking eyes are counterparts of our own, able to form images as ours do. If its eyes report that there is danger ahead, a squid can change color to match its surroundings. Ordinarily the Common Squid has red and purple dots scattered over the surface of its body. By expanding or contracting these pigment cells, it can vary its color from pearl white to dark brown. When cornered, a squid can squirt ink. The ink, secreted by a special ink-producing gland, provides a "smoke screen" and seems to paralyze a pursuer's sense of smell. Squid ink has, for centuries, been the source of India ink.*

In the waters of the Outer Lands, squid usually mate in the spring. The female fastens her fertilized eggs to rocks or shells. Covered with jelly, each pencil-shaped egg cluster contains about a hundred eggs. As the transparent jelly swells in the water, the developing embryos can be easily recognized. Hatching after a few weeks, newborn squid are brightly-colored active swimmers like their parents.

*Spanish restaurants serve "Squid in Ink," a dish of savory aroma and surprisingly attractive appearance.

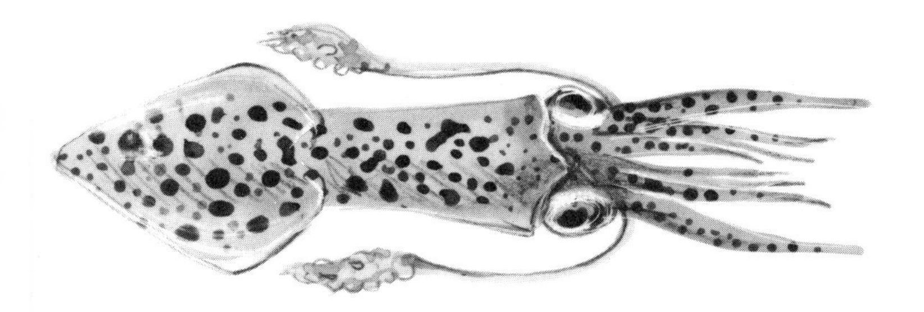

Common Terns

BIRDS

Gulls and terns are the most familiar birds of the bays and sounds. Members of the same sea-bird family, they nest in colonies on small offshore islands and on isolated barrier beaches. Their nests are built on the ground and their young begin to visit our beaches by mid-summer. Equipped with long narrow wings and webbed feet, these sea birds are superb fliers as well as swimmers, often ranging twenty-five miles a day for food.

Smaller than the gulls, the graceful terns have forked tails and pointed bills. Common Terns (Plate 9) hover over the water, diving from a height of ten feet or more when they see fish. Sometimes they submerge completely, surfacing moments later with Sand Eels* or Pipefish in their bills. When mackerel or bluefish chase smaller prey to the surface, the terns gather quickly. Fishermen who follow the flocking birds to find good casting grounds call them "Mackerel Gulls."

*A Sand Eel—or sand launce—is not an eel but a small silvery fish. Schools of them swim in shallow water during the summer, wriggling under the sand when they are chased. Fast and slippery, they are raked out of the sand by fishermen who use them for bait.

The noisy Herring Gulls (Plate 9) are scavengers. Occasionally they dive for fish. More often they cruise just above the surface of the water, watching for bits of refuse. At low tide they patrol the flats, boldly asserting their right to all the food along the shore. They feast on stranded Skates, crabs in tide pools, and clams and snails.

Unable to crack the heavy shell of a Moon Snail, a gull flies upward with its catch, drops it and zooms down after it. When the shell shatters, the gull picks out the meat. This is not reasoned behavior on the bird's part. The gull cannot tell the difference between hard and soft surfaces and may drop a snail over and over again on the same stretch of soft sand without breaking it open. Paved roads and parking lots, however, have proved a boon to the gulls, who litter highways with shells and sometimes use the flat roofs of shore cottages as targets.

Herring Gull

Herring Gulls were once considered rare birds in the Outer Lands. Nesting only on islands off the coast of Maine and further north, they were slaughtered by the thousands each year. Local people ate their eggs and chicks while hunters sold their pelts to the millinery trade. In the 1890s many a fashionable woman wore gull feathers on her hat. (Tern feathers were also in style and Roseate and Least Terns were almost exterminated.) After the passage of the first conservation laws at the turn of the century, the gulls began to make a comeback. Doubling, tripling, quadrupling their numbers, they have established breeding colonies on islands and sandbars as far south as New Jersey.

Today Herring Gulls are in the midst of a population explosion that parallels our own—and conservationists are considering ways to control them. Unlike other wild creatures whose existence is threatened by the growth of cities, they thrive in association with man. More people mean more garbage. More garbage means more gulls. During the summer they commute between the beaches, the fish piers and the town dumps. They freeload at waterfront restaurants, catching pieces of bread in mid-air or taking them from the tourist's outstretched hand.

Their population boom has created a serious problem at airports, where flocks roost on the runways, particularly during the winter. They are also driving other sea birds from nesting areas. The big Black-backed Gulls (Plate 9) can easily hold their own, but Laughing Gulls (Plate 9), Black Skimmers (Plate 9) and terns have been evicted from several islands on which they used to breed. So far the dispossessed terns are finding new nesting sites, but their future is in jeopardy unless protective measures are continued.

Shore birds such as the sandpipers and plovers are able to coexist with the gulls because their nesting habits and food tastes are so different. Of the dozen or so species of sandpipers and plovers that visit the Outer Lands only Piping Plovers and Spotted Sandpipers regularly nest there. The others visit the bay

Greater Yellowlegs
Semi-palmated Sandpipers
Semi-palmated Plover
Piping Plover
Spotted Sandpiper
Pectoral Sandpiper
Ruddy Turnstone

beaches in the spring and then fly to the Far North to breed. In July and August they return to the bays for a few weeks before migrating further south.

Long-legged, with webless feet, the shore birds hunt along the lines of sea wrack or wade in shallow water. Sometimes they dig up Horseshoe Crab eggs, but their usual diet consists of insects, worms and small crustaceans. Sandpipers (and their relatives, the Snipes and Yellowlegs) have long bills. Plovers are chunkier birds, with short stout bills. Turnstones, also in the Plover family, do just as their name suggests. They flip over stones and empty shells to find whatever tiny creatures may be hiding underneath.

Sea Rocket

PLANTS OF THE BEACH

Each tide washes up the remains of sea animals and weeds and each storm scatters sand over this debris. By fall the summer accumulation of driftwood, seaweed, fish skeletons, bird droppings has almost disappeared from view. Like fallen leaves in the woods, the buried beach litter decays, forming pockets of soil. Here and there in the sandy soil of the upper beach, a few plants manage to grow.

Baked by the sun, blown by the wind, showered with salt spray, these plants are stunted and scraggly. Although they are only a few yards from the sea, they cannot grow in salt water. Instead, they resemble desert plants. Like the cactuses of the desert, they are known as *xerophytes*, from the Greek meaning "dry plants." They send their roots deep into the sand for moisture, while their leaves and stems have developed different ways of conserving it.

Sea Rockets are often the only plants on a beach. Their fleshy leaves and stems store water as cactuses do. When their seeds ripen, the rocket-shaped seed pods roll down the beach to the bay. Washing ashore again, the pods are covered by drifting sand. In the spring the buried seeds start to grow. An annual with pale purple flowers, the Sea Rocket is a member of the Mustard family. Because its leaves and stems, buds and pods have a sharp mustardy taste, it is sometimes used in salads.

The little Seabeach Sandwort also has fleshy leaves. Its stalks are only a few inches high, but its roots may go three feet below ground. Related to chickweed, Seabeach Sandwort has small white flowers.

Seaside Spurge protects itself against dehydrating winds by hugging the sand. Instead of growing upright, its leathery leaves and sprawling red stems form a decorative mat on the beach. Like all the spurges, its roots and stems contain a sticky, milky juice.

The woolly gray covering on the leaves of Beach Wormwood—which you can rub off with your fingers—helps to prevent the evaporation of water. Known also as Dusty Miller,* Beach Wormwood is sometimes grown in gardens. Although its spreading roots make it difficult to transplant, it can easily be started from cuttings. Tall Wormwood, with its feathery, fernlike leaves, grows further back on the beach and in sandy soil inland. Both wormwoods are related to the Sagebrush of the western plains and to a European plant used in the manufacture of vermouth.

*At least four other plants are named Dusty Miller. All have gray-green leaves that look as if they had been dusted with flour.

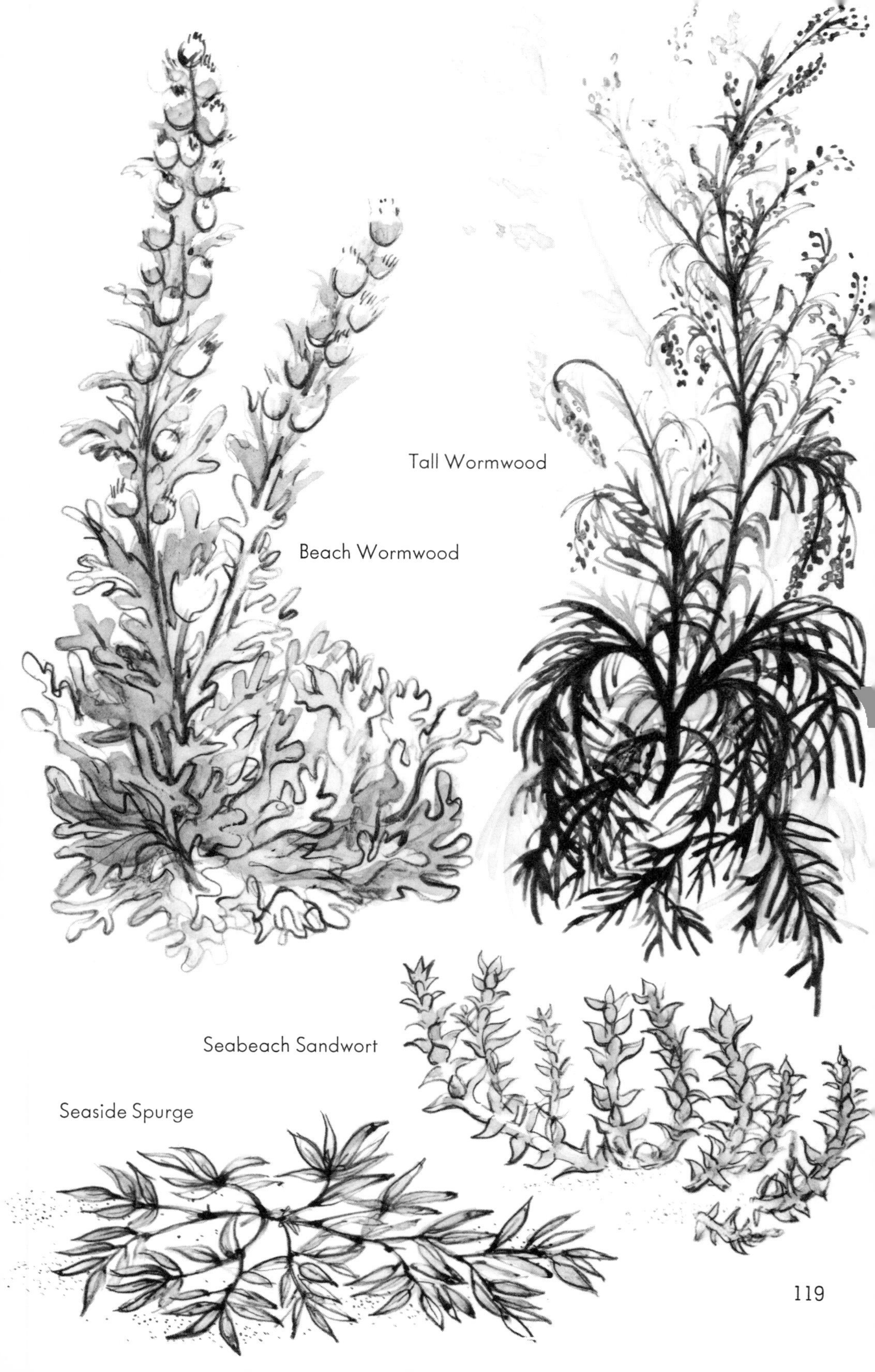
Tall Wormwood
Beach Wormwood
Seabeach Sandwort
Seaside Spurge

Short hairs on the leaves of Beach Clotbur also reduce water loss. The Clotbur's greenish flowers are inconspicuous, but its prickly burs are unmistakable. The burs, which contain the plant's seeds, stick to clothing and animal fur. Other species of Clotbur (which is also called Cocklebur) are common roadside weeds.

Beach Clotbur

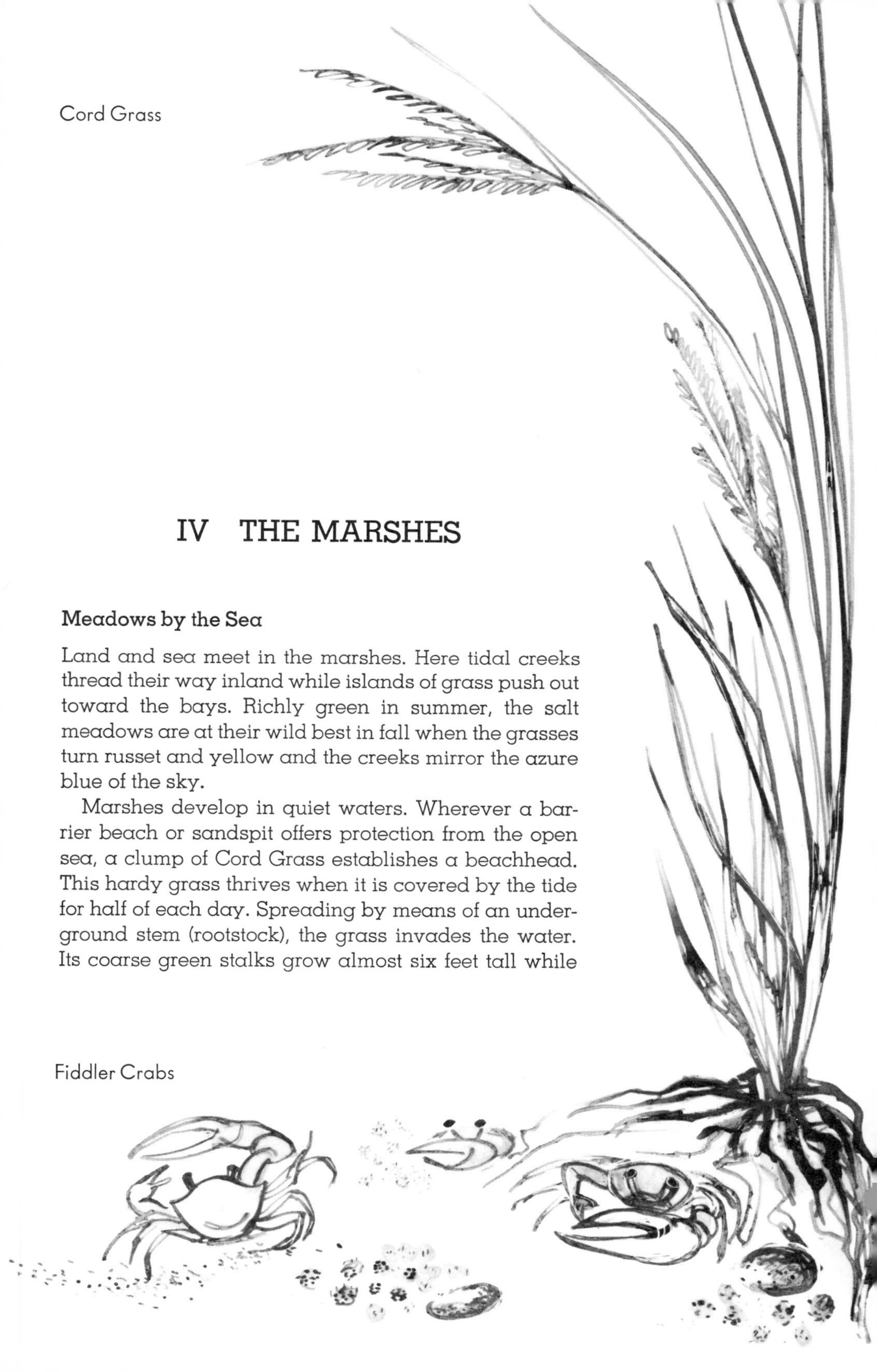

IV THE MARSHES

Meadows by the Sea

Land and sea meet in the marshes. Here tidal creeks thread their way inland while islands of grass push out toward the bays. Richly green in summer, the salt meadows are at their wild best in fall when the grasses turn russet and yellow and the creeks mirror the azure blue of the sky.

Marshes develop in quiet waters. Wherever a barrier beach or sandspit offers protection from the open sea, a clump of Cord Grass establishes a beachhead. This hardy grass thrives when it is covered by the tide for half of each day. Spreading by means of an underground stem (rootstock), the grass invades the water. Its coarse green stalks grow almost six feet tall while

Black Grass
Spike Grass
Salt-meadow Grass

its roots and stems intertwine on the muddy bottom of the cove. Dead leaves and sea sediment pile on top of the matted roots. Season after season, the clumps of Cord Grass widen. Year after year the debris around their roots builds up. Decaying slowly in the shallow water, it forms spongy hummocks of peat. Gradually, the floor of the cove rises, becoming too high and dry for Cord Grass. Then other grasses take over.

Salt-meadow Grass, which gets a salt-water bath only at high tides, is the dominant grass in most marshes. Growing vigorously, its slender leaves and wiry stems cover vast stretches of wetlands. Two feet tall, its stalks are often shaped by the wind into ragged "cowlicks"—flattened patches that look as if they have been trampled down. Salt-meadow Grass was once a valuable crop, supplying settlers with winter feed and bedding for their cattle and thatch for the roofs of their houses. When farmers led horses and carts through the mud to harvest tons of salt hay, the marshes were known as "the hay grounds."

Black Grass and Spike Grass grow along the high edges of the marsh. Black Grass, which was also harvested for salt hay, is not a grass but a rush. Its name comes from its dark green, pointed stalks and near-black flowers.

Over the centuries, these sturdy marsh grasses and the debris that accumulates around them may fill in an entire cove, building up the ground until it becomes dry land. This is what happened in tidewater Virginia where salt marshes were transformed into fertile farmlands. Or waves may break through the barrier beach, reclaiming the cove for the sea. More often, sea and marsh strike a balance. The grasses spread, their roots reaching deeper, their peat foundations thickening. But the level of the sea keeps pace with the rising islands of green. Flowing into narrow channels in the marshes, tidal waters eat away the edges of the grass-covered hummocks, keeping them in check. This equilibrium between land and sea can be seen in the thousands of acres of marsh along Long Island's south shore and in Cape Cod's Great Marshes between Sandwich and Barnstable. In the Great Marshes peat beds have been found twenty-three feet below the present high-water level. Carbon-14 dating of samples of this peat establishes its age at about 3660 years.

The Creeks

At high tide the marsh creeks flood, depositing silt in the tangled grass thickets. As the tide turns they carry organic matter—decaying grass stalks, fragments of peat, droppings of birds and animals—back to the bays. These products of the marshes enrich the offshore waters, supplying planktonic creatures and the teeming young of scallops and fish, oysters and clams with food. "All flesh is grass," the Bible says. Without the marsh grasses whose nutrient-rich remains are flushed into the bays, the fertility of our coastal waters would drop alarmingly—and the annual harvest of fish and shellfish might be cut in half.

Diamondback Terrapin

In addition to serving as a pantry for plankton, the fingers of bright blue water that wind around the grassy islands are refuges for lobsters in their soft shells, nurseries and feeding grounds for crabs and fish. Blue Crabs swim in and out with the tides. Striped bass, bluefish and menhaden spawn in the sheltered inlets. Young flounders spend a year or two hunting shrimp on the muddy creek bottoms before venturing out to the ocean.

Diamondback Terrapin live out their lives in the marshes. Paddling in the sluggish waters, they catch small fish and crabs, mussels and snails. In spring, female terrapins clamber up the slippery banks. Above the high-tide line, they scoop out deep holes in which to lay their eggs. Hatching in late summer, the baby terrapins remain in their nests until the following year.

Sometimes a brook, following a channel carved by an ancient meltwater stream, makes its way to the sea through the marshes. Then the creeks become part of a chain of waterways, linking bay with upland pond. Salt and fresh water mingle. The salt water, which is heavier, moves along the bottom of the channel with the fresh water flowing over it. When this happens, the marsh becomes a passageway for migrating fish and a modest creek is likely to be dignified with the name of "Herring River."

See How They Run

"There are many Herring Rivers on the Cape," Thoreau wrote more than a century ago. "They will perhaps be more numerous than herrings soon." Although the herring population has declined precipitously, there are still places on the Cape and islands where their annual migration can be seen.

Each spring schools of big-eyed, silver-sided fish assemble in the bays. These are the herring, more correctly called Alewives to distinguish them from their larger relatives the sea herring. Ocean dwellers for most of their lives, they spawn in fresh water—not in any convenient stream, but usually in the same brook or pond where they themselves hatched several years earlier.

From March until June, the Alewives swim into the marshes. Crowding into the creeks, they fill them from bank to bank. As they travel inland, they struggle against the current and leap up fish ladders until they reach their spawning place. In early summer, the fish make their return journey through the marshes to the sea, followed a few weeks later by their finger-long young.

From the days of the Pilgrims, the herring run has been a time of high excitement along the shore. Netted by the thousands, the fish were fried, smoked or salted in tubs of brine, while any surplus went to fertilize hills of corn. The catch was so valuable that when a new dam in Falmouth blocked the passage of the fish a Herring Party and an Anti-Herring Party formed. During a protracted fight the Anti-Herrings filled a cannon on the village green with Alewives and fired it off, blowing fish, cannon and the man who fired it to bits.

Even though modern tastes favor a less bony fish there is still a holiday air on the Cape and islands during the herring run. Gulls cover the marshes. Screaming, wheeling, diving, they proclaim the arrival of the fish and trail them upstream. Flocks of Blue Jays and grackles follow to see what all the excitement is about. Farther along the route of the run, people congregate—children brought by school bus, men with dip nets, women with cameras and picnic baskets—all stirred by the spectacle of these deep-sea creatures who travel hundreds of miles each year to lay their eggs in the fresh-water ponds where they were spawned.

Less of an annual spectacle, but equally remarkable, are the journeys of the eels who migrate through the marshes in the opposite direction. These serpent-like fish spawn in the warm salty waters of the Sargasso Sea and their young are carried north by the Gulf Stream. Yearling eels, two or three inches long

Herring run

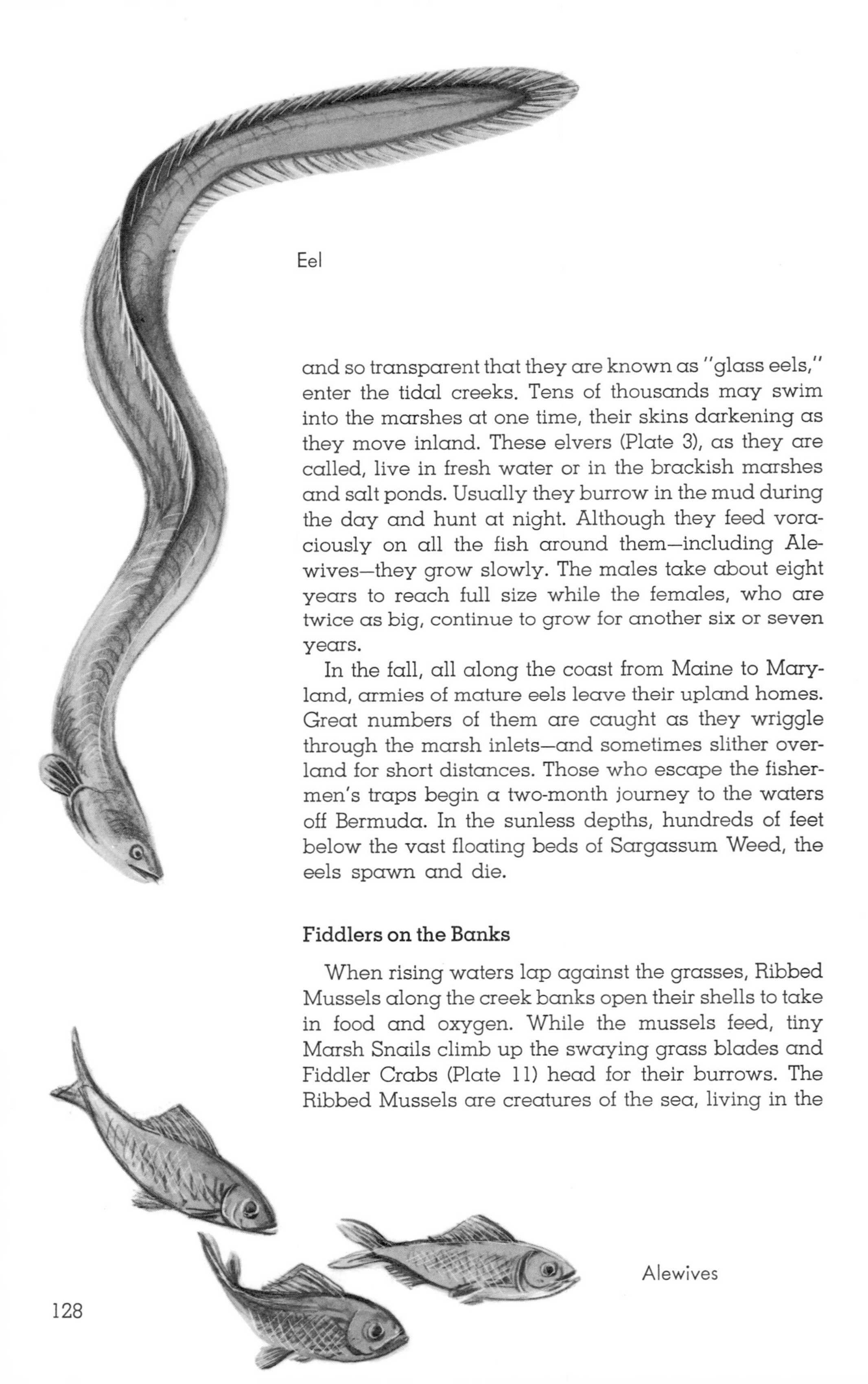

Eel

and so transparent that they are known as "glass eels," enter the tidal creeks. Tens of thousands may swim into the marshes at one time, their skins darkening as they move inland. These elvers (Plate 3), as they are called, live in fresh water or in the brackish marshes and salt ponds. Usually they burrow in the mud during the day and hunt at night. Although they feed voraciously on all the fish around them—including Alewives—they grow slowly. The males take about eight years to reach full size while the females, who are twice as big, continue to grow for another six or seven years.

In the fall, all along the coast from Maine to Maryland, armies of mature eels leave their upland homes. Great numbers of them are caught as they wriggle through the marsh inlets—and sometimes slither overland for short distances. Those who escape the fishermen's traps begin a two-month journey to the waters off Bermuda. In the sunless depths, hundreds of feet below the vast floating beds of Sargassum Weed, the eels spawn and die.

Fiddlers on the Banks

When rising waters lap against the grasses, Ribbed Mussels along the creek banks open their shells to take in food and oxygen. While the mussels feed, tiny Marsh Snails climb up the swaying grass blades and Fiddler Crabs (Plate 11) head for their burrows. The Ribbed Mussels are creatures of the sea, living in the

Alewives

same way as Edible Mussels. After hatching in the creeks, they anchor themselves to the base of the grasses, forming thick beds in the soggy mud.

The Marsh Snails—sometimes called Coffee-beans because of their shiny brown shells—are built like land snails. Equipped with lungs rather than gills, they climb out of the water to breathe.

In the half-land, half-sea world of the marshes, Fiddler Crabs are half-way between. They breathe in the water as their Swimming Crab relatives do, but they can store oxygen in their gill chambers and live on land for weeks at a time. At low tide, thousands of these lively, scrappy little animals scurry through the grass thickets, feeding on minute plants which they scrape off the surface of the mud. They turn their stalked eyes warily as they eat, alert for the shadow of a gull or the long-billed Whimbrels that feast on them.

The Fiddlers of the East Coast travel in herds, moving as a unit the way Sanderlings do. Although they lack ears, they are quick to pick up the vibrations of a footstep. If you advance toward them suddenly, you can drive them ahead of you like sheep. Soon every Fiddler on the marsh will disappear, taking cover in its burrow until all is quiet again.

Three species of Fiddlers live in the Outer Lands, but they resemble each other closely in size and in the appearance of their mottled polished shells. Only males have the oversized claw which is their trademark. All young Fiddlers start out with paired claws of equal size, but the male's right claw, growing more rapidly than the rest of its body, becomes enormous.

In the mating season, the male raises his big claw at the approach of a female and waves it back and forth like a signal flag. During the rest of the year he uses his oversized claw for defense. He sidles out of his burrow claw first, as if spoiling for a fight. When he crawls back in, the big claw protects his rear, its pincers opening and closing threateningly. (A Whimbrel, thrusting its curved bill into a burrow, finds the big claw a convenient handle by which to seize the Fiddler.) If the claw is lost in battle a new big one appears at the next molt, only this time it grows on the crab's left side. Often half of the males in a Fiddler colony are left-clawed.

After mating, a female Fiddler carries her eggs on her abdomen. She crawls down to the creek at hatch-

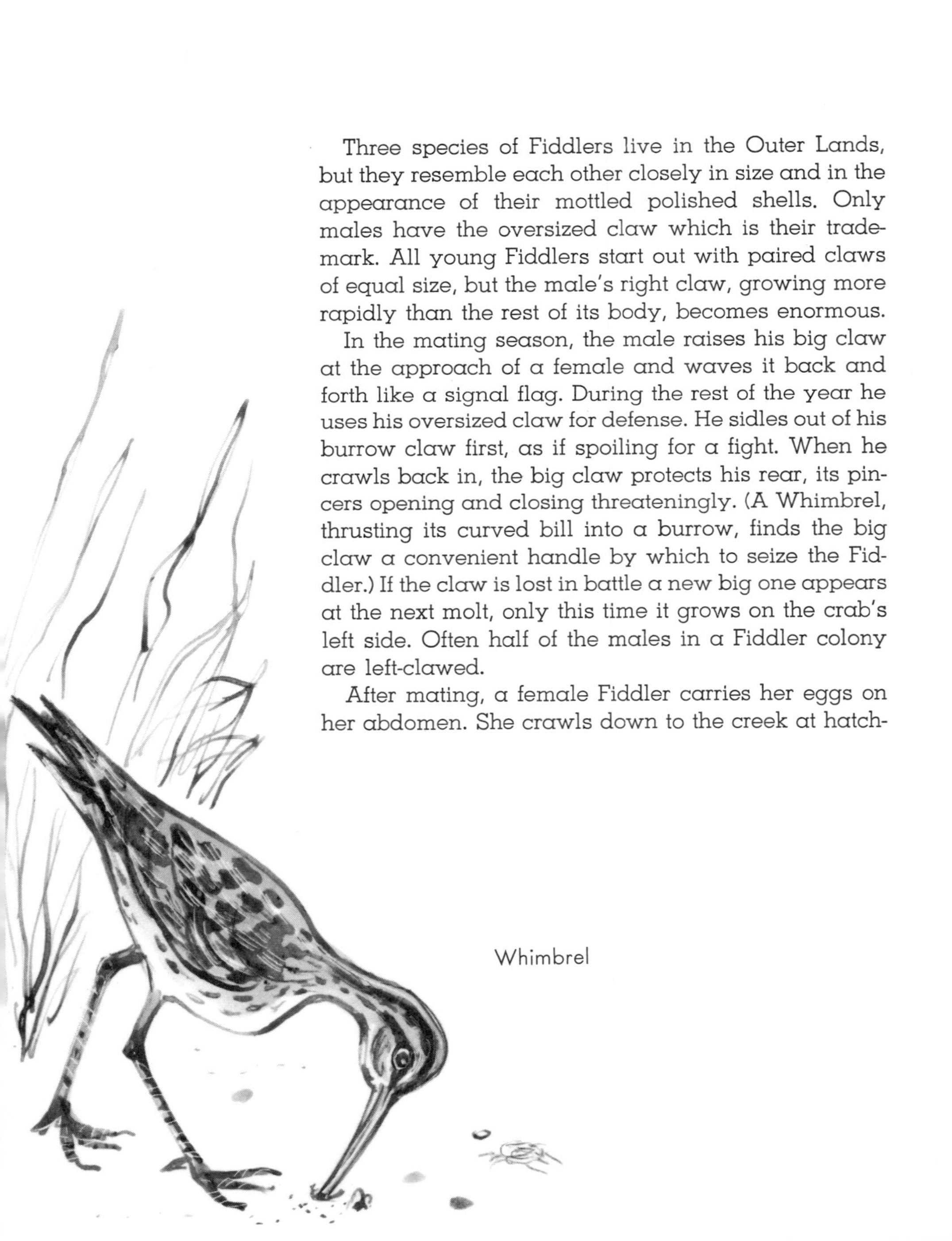

Whimbrel

Fiddler Crab

ing time to release her young in the water. The newly hatched Fiddlers swim in the creek for a few weeks before returning to its banks.

The Fiddlers' burrows are slanting tunnels about three feet long, with a room at the far end. The crabs dig with their walking legs, shaping the mud they excavate into balls and carrying it outside. At high tide they plug up the tunnel entrances so that their rooms remain damp but watertight. The crabs' constant working and reworking of the mud helps to aerate the waterlogged soil and speeds up the decay of organic matter. The mud that they dig out of their burrows is washed away, adding to the food supply of the coastal waters.

Active night and day depending on the rhythm of the tides, Fiddlers go through a remarkable series of color changes. At dawn their shiny shells begin to darken, growing darkest at low tide, as if to camouflage them when they scuttle across the gray marsh mud. At dusk their shells become lighter—and lighter still at high tide when the crabs are in their burrows.

The source of these color changes is no mystery. Along with many other marine creatures, Fiddlers have spots of pigment in their shells which can spread out or contract. But what triggers the movement of the pigment grains?

Unlike a squid whose color changes depend on messages from its eyes, a Fiddler seems to be governed by internal clocks, one running on sun time and the other following the tides, which are controlled by the moon. When kept in constant darkness in a labora-

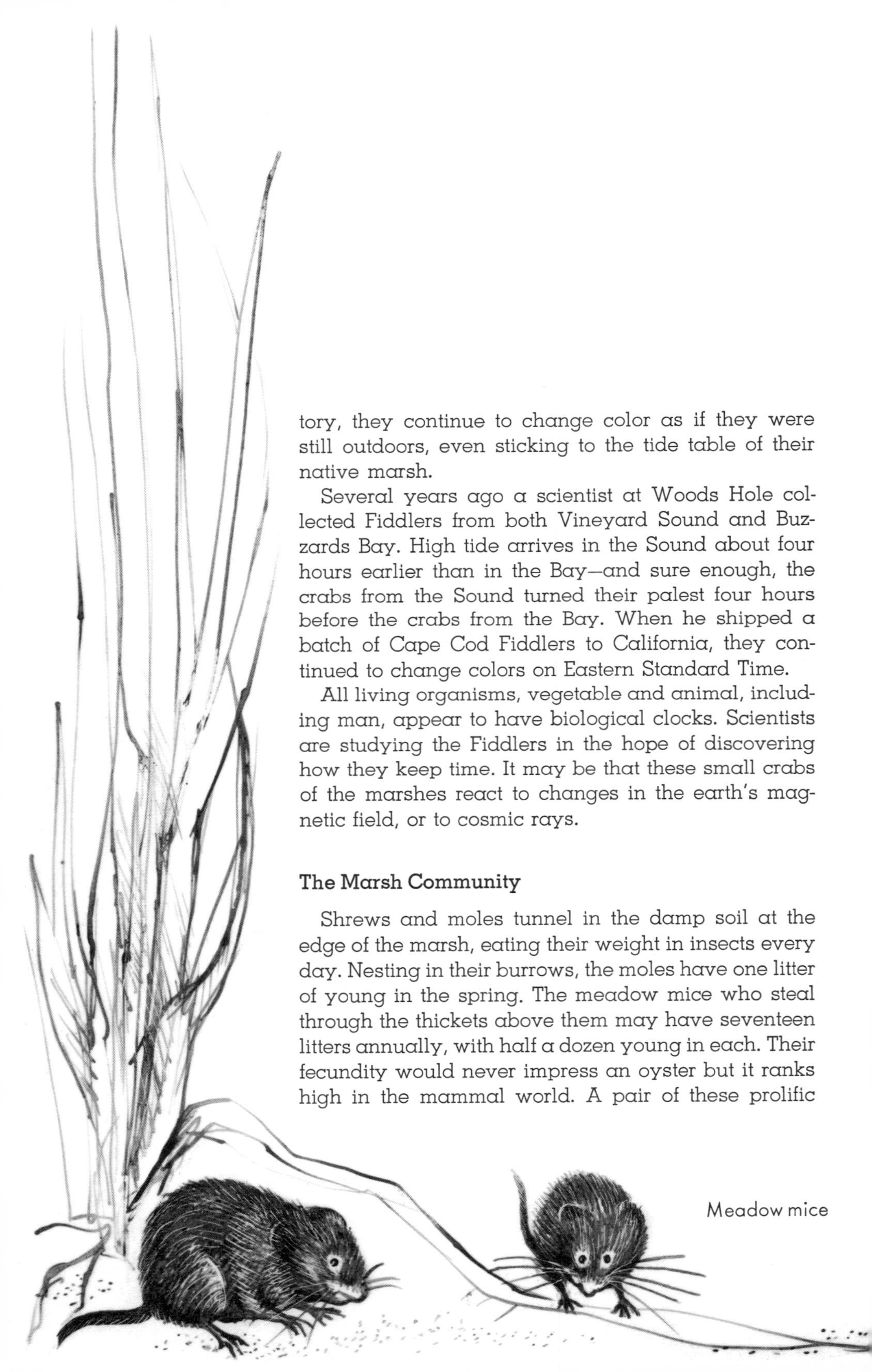

tory, they continue to change color as if they were still outdoors, even sticking to the tide table of their native marsh.

Several years ago a scientist at Woods Hole collected Fiddlers from both Vineyard Sound and Buzzards Bay. High tide arrives in the Sound about four hours earlier than in the Bay—and sure enough, the crabs from the Sound turned their palest four hours before the crabs from the Bay. When he shipped a batch of Cape Cod Fiddlers to California, they continued to change colors on Eastern Standard Time.

All living organisms, vegetable and animal, including man, appear to have biological clocks. Scientists are studying the Fiddlers in the hope of discovering how they keep time. It may be that these small crabs of the marshes react to changes in the earth's magnetic field, or to cosmic rays.

The Marsh Community

Shrews and moles tunnel in the damp soil at the edge of the marsh, eating their weight in insects every day. Nesting in their burrows, the moles have one litter of young in the spring. The meadow mice who steal through the thickets above them may have seventeen litters annually, with half a dozen young in each. Their fecundity would never impress an oyster but it ranks high in the mammal world. A pair of these prolific

Meadow mice

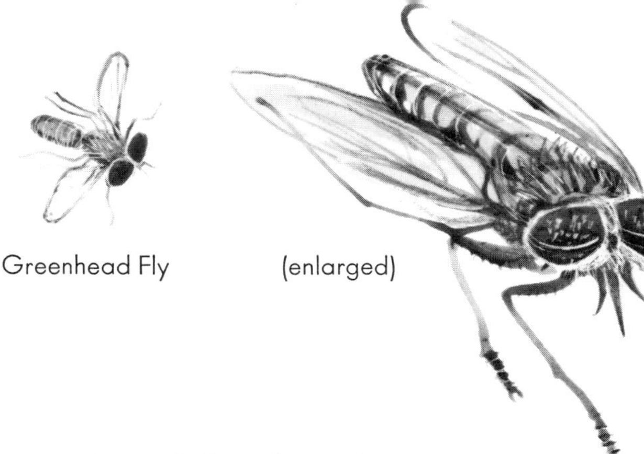

Greenhead Fly (enlarged)

creatures could have more than a billion descendants in a year if enemies failed to keep them in check. These sleek short-tailed mice cut winding runways through the jungle of grasses by snipping off the stems with their sharp teeth. They eat prodigiously, a single mouse consuming about thirty pounds of vegetation a year.

Fireflies flash their lights across the marsh at night while their young hunt snails. Beetle grubs chew on the roots of the grasses, Salt-marsh Caterpillars nibble on the stalks, and Salt-marsh Mosquitoes buzz around the waving blades. The mosquitoes lay their eggs in tide pools and puddles on the marsh. Their wriggling young live in the water, feeding on microscopic plants and animals and breathing air through snorkel-like tubes. After about a fortnight, the wrigglers change form, becoming first big-headed pupae and then winged adults who take off from the surface of the water. The inoffensive male mosquito feeds on nectar and plant juices, but the female needs an occasional sip of mammal blood before her eggs can ripen.

The army of Greenhead Flies that invades the beaches for a week or two in mid-summer is also based in the marshes. The female Greenhead, another blood-drinker, lays her eggs on stalks of Cord Grass. Compared to the mosquitoes, the flies' life cycle is a long one. Although the adults die soon after egg-laying, their wormlike young live in the mud, feeding on insects and snails until the following summer. When the

Salt-marsh Caterpillar

Tree Swallows

marsh soil is warm, they become adults. Normally this transformation takes place in mid-July, but if the summer has been cold and rainy, they remain in the ground for another twelve months. That is why the Greenhead population varies from year to year, reaching its peak when the days are sunny and hot.

Good weather means more Greenheads—and more birds. Public nuisances though they are, the Greenheads and mosquitoes are links in the complicated food-web that makes the marshes one of the richest areas along the coast for birds. Waterfowl and marsh birds, song birds and birds of prey all find shelter and food in the salt marshes.

In late spring Cord Grass stalks conceal the nests of Seaside Sparrows and Clapper Rails (Plate 10). Colonies of Sharp-tailed Sparrows (Plate 11) lay their eggs in thickets of Salt-meadow Grass, and Marsh Hawks (Plate 10) nest along the dry edges of the marsh, beyond the reach of the tides. Both groups of sparrows are ground-dwellers who glean grass seeds, insects and sandhoppers, the Seasides also adding crabs to their diet. Clapper Rails—sometimes called Salt-water Marsh Hens—feed along the creek banks, using their long bills to spear Fiddlers and open mussel shells. Patrolling the air above the salt meadow, Marsh Hawks carry off mice, young rabbits—and sparrows.

The nesting birds are joined by birds from the uplands—Robins and Barn Swallows collecting mud for their nests, and Kingbirds skimming over the tops of the grasses to catch insects. Tree Swallows are so val-

Sparrow Hawk

uable as insect-eaters that cranberry farmers put up nesting boxes on long poles to encourage the swallows to settle in their bogs. At dusk, Great Horned Owls swoop down on unwary animals while Black-crowned Night Herons (Plate 10) go fishing. The herons nest in bushes and trees near the water. Sleeping during the day, they catch fish, crabs and mice on their nightly rounds.

When the shore birds return from their northern nesting sites, Yellowlegs wade in the creeks, Pectoral Sandpipers hunt in the grass thickets and Whimbrels renew their claim to the banks where the Fiddlers live. As the summer progresses, Sparrow Hawks cruise above the marsh and Great Blue Herons visit the creeks. Scarcely larger than a Robin, a Sparrow Hawk hovers in mid-air with claws extended. Pouncing on a cricket or caterpillar, it carries its prey to a telephone wire to dine in leisurely fashion. The Great Blue Heron stalks Alewives and eels on their way back to the sea. Flapping its lordly wings, it takes off with deliberate speed if a spectator draws near.

Other visitors appear as the days grow shorter. From August through October, flights of southbound land and sea birds stop in the marshes to rest and feed. When fresh-water ponds freeze over, great flocks of ducks and geese and wintering swans paddle along the tidal creeks. The diving ducks eat Marsh Snails and mussels. Black Ducks (Plate 10) gather grass seeds, and the majestic Canada Geese (Plate 10) feed on the rootstocks of Cord Grass and other marsh plants.

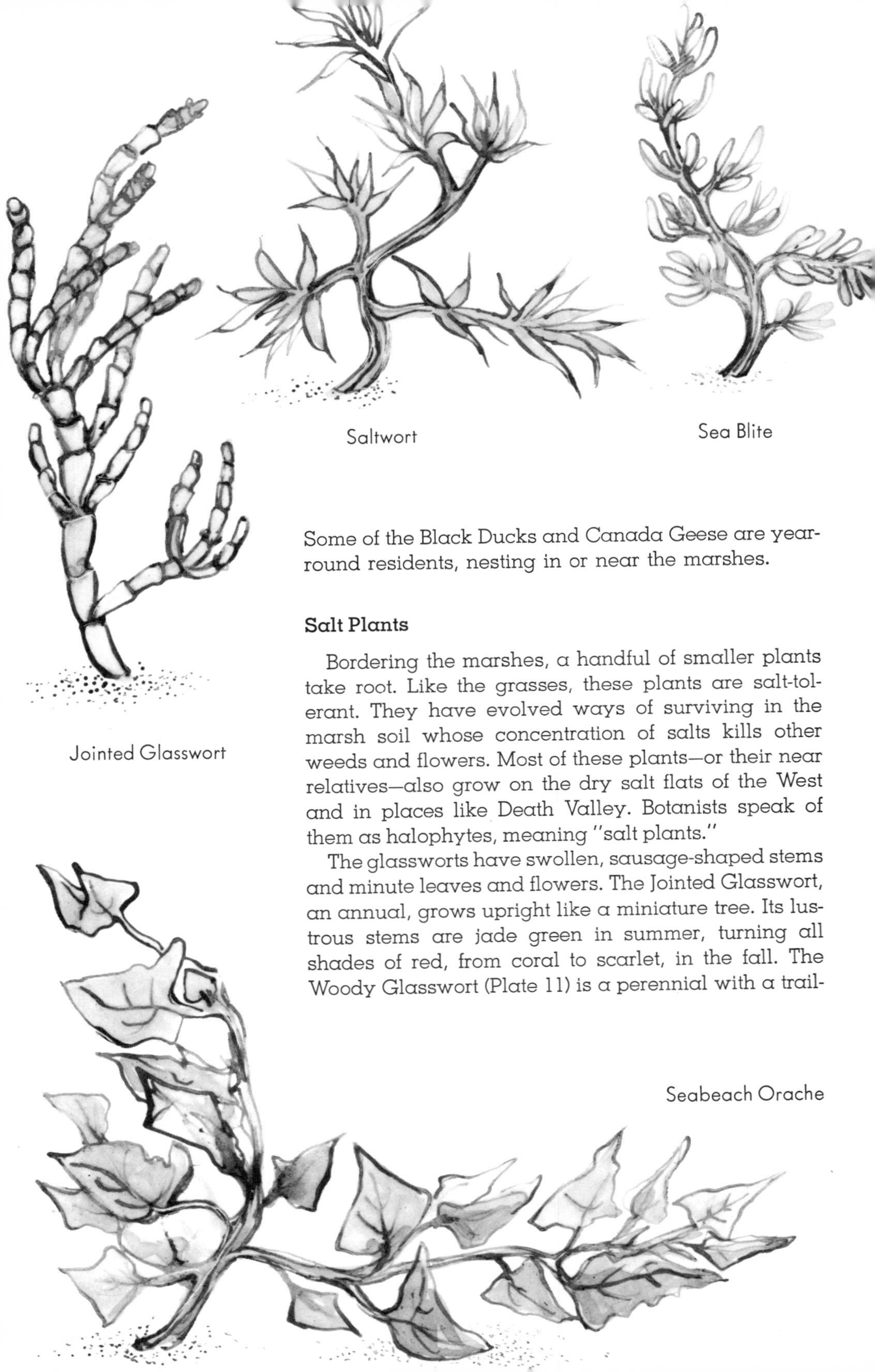

Some of the Black Ducks and Canada Geese are year-round residents, nesting in or near the marshes.

Salt Plants

Bordering the marshes, a handful of smaller plants take root. Like the grasses, these plants are salt-tolerant. They have evolved ways of surviving in the marsh soil whose concentration of salts kills other weeds and flowers. Most of these plants—or their near relatives—also grow on the dry salt flats of the West and in places like Death Valley. Botanists speak of them as halophytes, meaning "salt plants."

The glassworts have swollen, sausage-shaped stems and minute leaves and flowers. The Jointed Glasswort, an annual, grows upright like a miniature tree. Its lustrous stems are jade green in summer, turning all shades of red, from coral to scarlet, in the fall. The Woody Glasswort (Plate 11) is a perennial with a trailing main stem and rows of erect branches. Geese feed on the fleshy glasswort stems, ducks eat their tiny seeds and housewives along the coast use them for making pickles. This is why they are sometimes called Pickle Plants. But they are also known as Samphire, Chickens' Toes, Frog Grass and—in the West—Burroweed. Their scientific name, *Salicornia*, comes from the Greek, meaning "salt-horn." Like many seashore plants they have a pleasant salty taste.

Sea Blite and Saltwort (Plate 11) are both annuals with small succulent leaves and inconspicuous flowers. Although they look somewhat alike, Saltwort's dull green leaves have sharp spines at their tips. Seabeach Orache is the only one of the group whose leaves are not fleshy. Ashy-green and almost triangular, they can be cooked and eaten like spinach. Don't salt them, however, for their seasoning is built in. The glassworts, Sea Blite, Saltwort and Seabeach Orache all belong to the Goosefoot family, a group of plants that also includes spinach and beets.

The showy pink flowers of Rose Mallow and Seaside Gerardia (Plate 11) brighten the marshes in midsummer. Nearby, Sea Lavender's (Plate 11) leathery leaves lie on the ground while its bushy flower-stalk shoots up. Hundreds of tiny fragrant flowers open in late summer. Keeping their color through the fall, the flower clusters form a purple mist along the edge of the marsh. When winter winds break the heavy stalks, they roll across the ground like tumbleweeds, scattering their seeds.

Red-winged Blackbird on Cattail

The Fresh Marsh

Along the upper reaches of the creeks where fresh-water streams dilute the sea's salty influx, plant and animal life changes abruptly. In a fresh marsh, Cattails and Bulrushes take the place of grasses. Their branching rootstocks establish firm platforms in the mud while their leaves remain above water. Although the minute seeds that make up the Cattail's brown velvet spike are too small to attract birds, Bulrush seeds are eaten by ducks and other waterfowl. Colonies of Red-winged Blackbirds and Long-billed Marsh Wrens nest in the Cattail marshes and muskrats thrive in their waterways. The muskrats use the leaves of the plants for lining their lodges, and feed on the starchy rootstocks during the winter months.

As in a salt marsh, the accumulation of leaves and rootstocks tends to fill in the land. A meadow of Sedges follows the Cattails and Bulrushes and in time is replaced by woody plants—Marsh Elder, Sumac, Buttonbush, and the omnipresent Poison Ivy. These thickets, which shelter rabbits and deer and a variety of game birds and song birds, may eventually give way to Willows and Alders, Red Maples and Black Gum trees.

The Black Gum has more aliases than a Quahog. In

Long-billed Marsh Wren on Bulrush

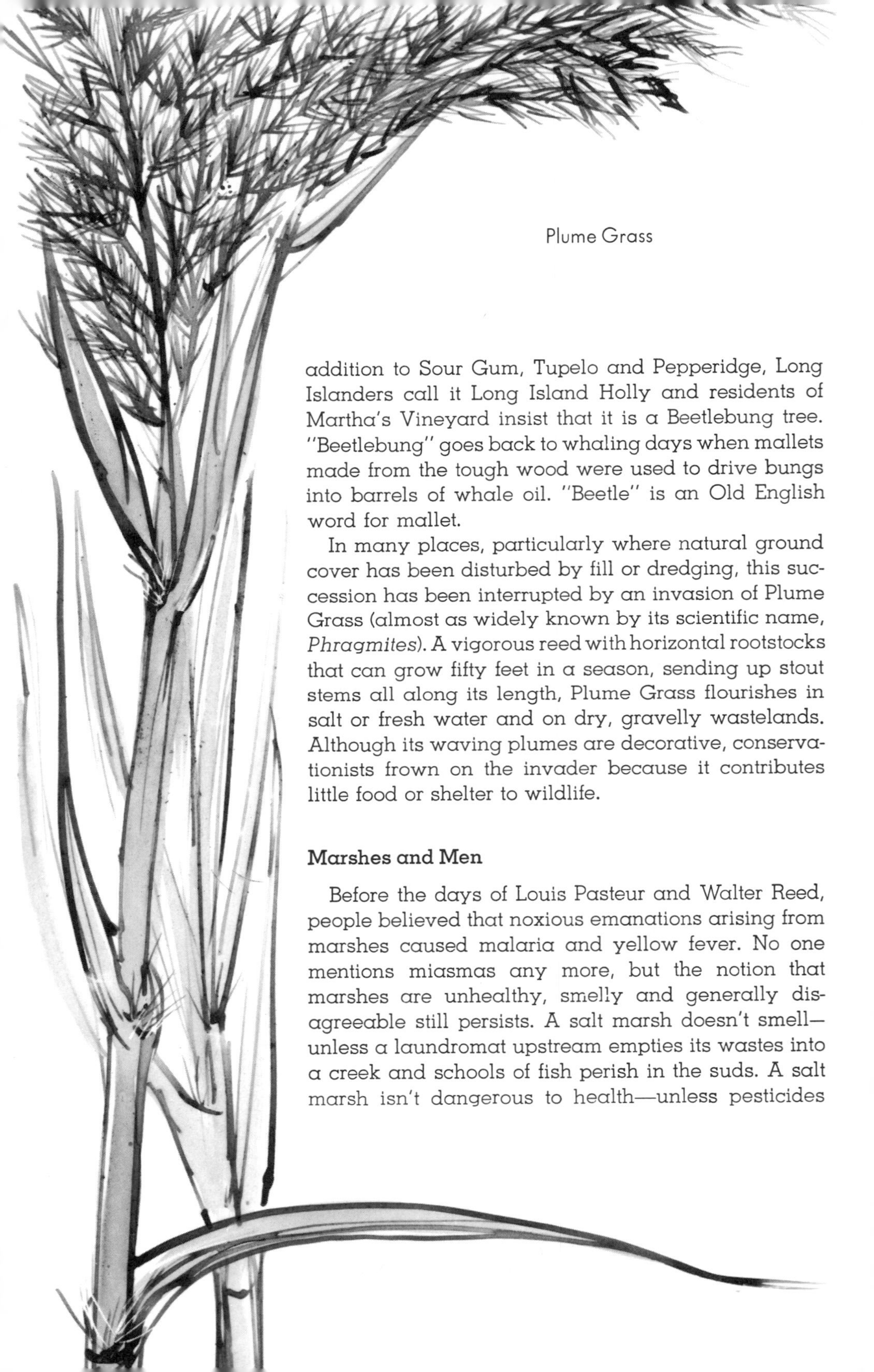
Plume Grass

addition to Sour Gum, Tupelo and Pepperidge, Long Islanders call it Long Island Holly and residents of Martha's Vineyard insist that it is a Beetlebung tree. "Beetlebung" goes back to whaling days when mallets made from the tough wood were used to drive bungs into barrels of whale oil. "Beetle" is an Old English word for mallet.

In many places, particularly where natural ground cover has been disturbed by fill or dredging, this succession has been interrupted by an invasion of Plume Grass (almost as widely known by its scientific name, *Phragmites*). A vigorous reed with horizontal rootstocks that can grow fifty feet in a season, sending up stout stems all along its length, Plume Grass flourishes in salt or fresh water and on dry, gravelly wastelands. Although its waving plumes are decorative, conservationists frown on the invader because it contributes little food or shelter to wildlife.

Marshes and Men

Before the days of Louis Pasteur and Walter Reed, people believed that noxious emanations arising from marshes caused malaria and yellow fever. No one mentions miasmas any more, but the notion that marshes are unhealthy, smelly and generally disagreeable still persists. A salt marsh doesn't smell—unless a laundromat upstream empties its wastes into a creek and schools of fish perish in the suds. A salt marsh isn't dangerous to health—unless pesticides

Beetlebung Corner, Martha's Vineyard

have been sprayed in the area. Then whole populations of Blue Crabs and Fiddlers, oysters and mussels may be wiped out. Although the pesticides which have replaced DDT break down fairly rapidly on land, several are persistent in sea water. Not only are they highly toxic to fish, but their poisonous residues build up in the tissues of shellfish. The amount of methoxychlor stored by an oyster is almost 6,000 times greater than the concentration of the poison in the surrounding water.

To avoid pesticides, many Outer Lands communities get rid of Salt-marsh mosquitoes by ditching marshy areas. The ditches drain standing water on the surface of the marsh where mosquitoes might breed, and also allow tidewater to flow in. With the tides come hungry minnows who snap up the mosquito wrigglers almost as fast as their mother can lay eggs. The rectangular wooden boxes which dot Cape Cod marshes offer a comparable control for Greenhead Flies. The greenheads enter the boxes from below and fly upward toward the light until they hit a screen which blocks their way. Unable to backtrack and fly out, they eventually die. No one is sure why these traps work, but they rid the marshes of millions of flies each summer without the use of poison sprays.

Salt marshes have more than negative virtues, however. As acre after acre of marshland has been dredged and filled, and mile after mile of waterway polluted, there has been an irremediable loss of fish and shellfish, animals and birds. Once oysters flourished at the mouths of marsh creeks and clams grew plump in the marsh mud. Now the oyster and clam harvests in the Outer Lands grow scantier each year. Lobsters—and even Clam Worms—are presently imported from Maine, and Blue Crabs from Maryland. Herds of Fiddlers have vanished from many areas. Diamondback Terrapins are rare and Night Heron rookeries which fifty years ago housed thousands of nesting birds now shelter only a few dozen.

Only the wintering populations of ducks and geese are on the increase, because of the wildlife refuges that have been established. These refuges, along with national and state parks and local nature centers, point to a future that has some hope. With public recognition of the value of coastal wetlands, state legislatures have begun to pass laws to preserve them. The clock cannot be turned back to 1620—or 1920—but there is still a chance to save some of the dynamic marsh communities and the interlocking lives that depend on them.

Beach Grass

V THE DUNES

Building a Dune

A steady sea wind sweeping across the beach carries grains of sand inland. When its motion is interrupted by a log or grass clump, the wind drops its burden of sand. Slowly, a mound builds up. Growing higher, broader, merging with other mounds, it becomes a hillock, a ridge—a dune. Rolling up the face of the dune and tumbling over its crest, the wind-blown sand gives the dune its characteristic shape—a gentle slope on the windward side and a sharp drop on the lee.

The dune may continue to rise, reaching a height of one hundred feet or more. Or a second dune may grow behind the foredune, with a broad sandy valley in between. If nothing impedes the wind or anchors the

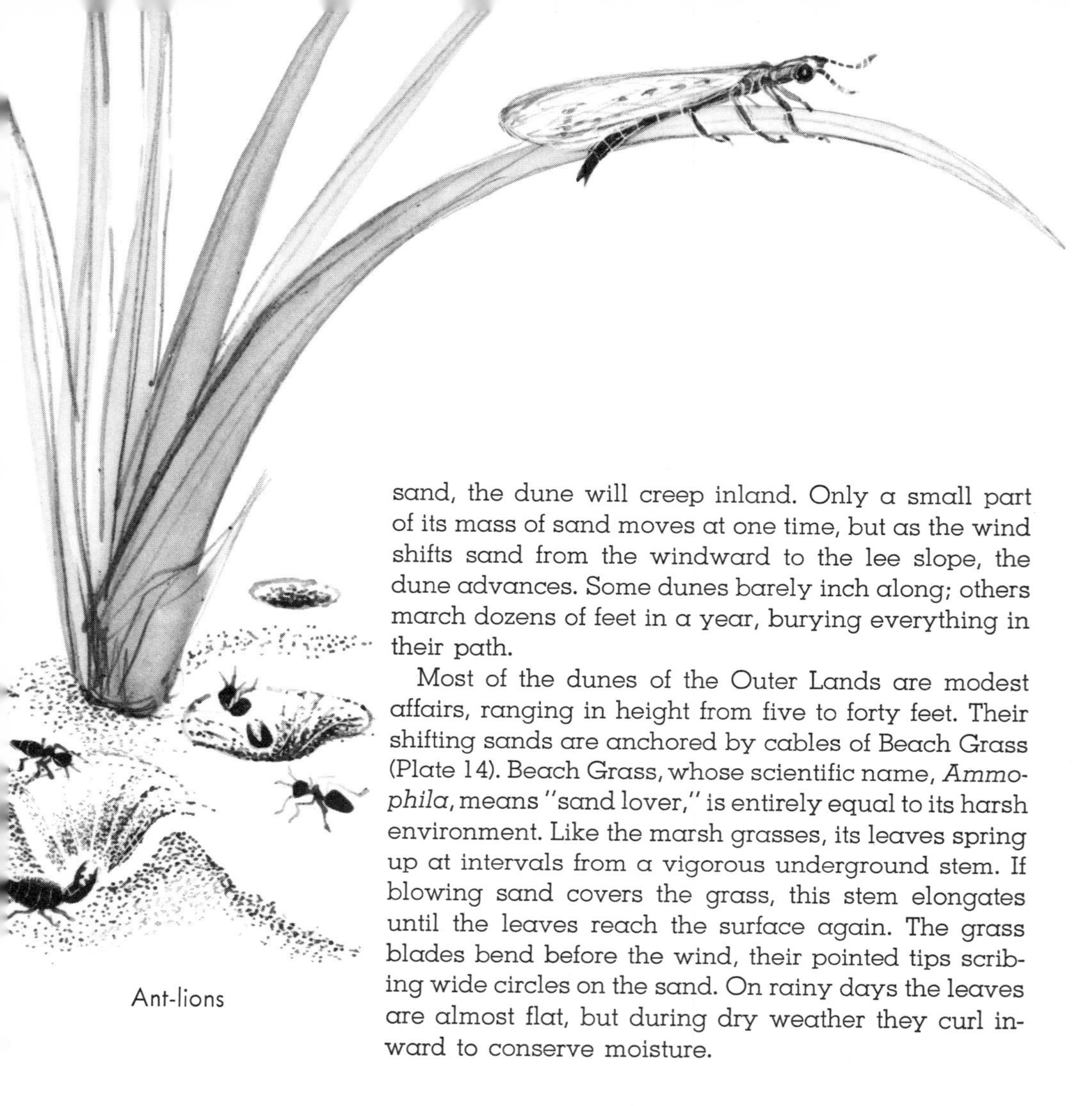

sand, the dune will creep inland. Only a small part of its mass of sand moves at one time, but as the wind shifts sand from the windward to the lee slope, the dune advances. Some dunes barely inch along; others march dozens of feet in a year, burying everything in their path.

Most of the dunes of the Outer Lands are modest affairs, ranging in height from five to forty feet. Their shifting sands are anchored by cables of Beach Grass (Plate 14). Beach Grass, whose scientific name, *Ammophila*, means "sand lover," is entirely equal to its harsh environment. Like the marsh grasses, its leaves spring up at intervals from a vigorous underground stem. If blowing sand covers the grass, this stem elongates until the leaves reach the surface again. The grass blades bend before the wind, their pointed tips scribing wide circles on the sand. On rainy days the leaves are almost flat, but during dry weather they curl inward to conserve moisture.

Ant-lions

Dune Neighborhoods

After clumps of Beach Grass have battened down the sand, other plants and animals can live there. Beach Peas (Plate 12) trail down the slopes of the foredune, their purple flowers producing crops of tiny edible peas all summer. Seaside Goldenrod (Plate 12) takes root alongside them. Unlike its inland relatives, this goldenrod of the dunes is a succulent, able to store

Wolf Spider

water in its thick, knobby stalk. A late bloomer, it flowers from August until the first frost.

With the surface temperature of the foredune soaring to 120 degrees in summer, most of its animal residents conceal themselves during the day and forage in the cooler evening hours. A neat round hole near a Beach Pea vine marks the entrance of a Wolf Spider's burrow. The long-legged Wolf Spiders are hunters who run down their victims instead of trapping them in webs. Only a female spider digs a burrow, remaining underground during the summer to raise a brood of young. Her smaller mate finds shade under a leaf and comes out at night to hunt.

Beach Plum

Ant-lions dig cone-shaped craters by backing into the sand and using their heads as shovels. When the excavation has been completed the Ant-lion conceals itself in the loose sand at the bottom of its pit and waits for an ant or other small prey to pass by. As the ant flounders down the steep slopes of the crater, the Ant-lion's pincer jaws shoot up to grab it. An Ant-lion may live for two years in its pit on the dune, spinning a cocoon under the ground after it is full-grown. Weeks, sometimes months, later, a gauzy-winged insect emerges from the cocoon. Adult Ant-lions live only long enough to mate and lay eggs on the sand.

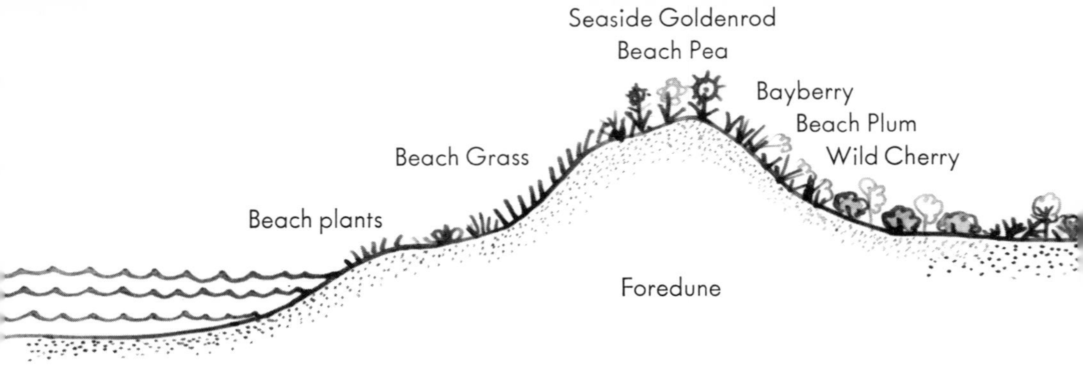

Dune neighborhoods

Feeding on the coarse leaves of Beach Grass, Seaside Grasshoppers venture outdoors during the day—but they have their own system of air-conditioning. When the sand is hot, they stretch their long legs and raise their bodies above it. If it becomes still hotter, they fly, leaping into a layer of cooler air above the dune. Only when they leap with a noisy whirring of wings do the grasshoppers make their presence known. At rest on the dune, their ivory-tinted bodies match the sand so perfectly that they are virtually invisible.

As the foredune dips down, woody shrubs—Bayberry (Plate 13) and rose, Beach Plum, Wild Cherry, Poison Ivy—grow in the valley behind it. With their roots close to the water table and their branches receiving some protection from wind and salt spray, these shrubs are able to form dense thickets.

Bayberries (also known as Wax Myrtle) have stiff gray branches and glossy oval leaves. Because their male and female flowers are borne on separate bushes, only female plants carry the clusters of gray fruit from which fragrant bayberry candles are made. When the berries are boiled, their waxy coating melts and floats on the surface of the water. After straining, the wax is remelted and poured into candle molds. Commercial bayberry candles are often dyed, but their natural color is gray-green. The leaves are also aromatic. They are not the commercial bay leaves—which come from Bay Rum trees in the tropics—but they add a spicy flavor of their own to soups and stews.

Seaside Grasshopper

Secondary dune

The sturdy Salt-spray Rose (Plate 12) seems at home on the dunes, though its ancestors arrived from China little more than a century ago, floating ashore from a sailing vessel wrecked off Cape Cod. Thornier and with coarser leaves than native wild roses, it has pink or white flowers. The plump red rose hips are filled with bony seeds.

Beach Plums (Plate 13) and Wild Cherries are also members of the Rose family. Closely related plants, their flowers open at almost the same time in spring and their purple fruits ripen in the fall. The juicy plums are picked to make jelly, while the astringent cherries are usually left to the birds. Beach Plums grow only in the sandy soil of dunes and moors, but Wild Cherries become tall handsome trees when they move inland.

Poison Ivy, whose three shiny leaflets spell "Keep away!", has many disguises. On a bare dune, it is a slender vine like the Beach Pea. In a protected hollow it becomes a shrub with an upright trunk and spreading branches. In the woods it is a vigorous climber, fastening itself to the bark of a tree with clinging aerial roots. Poison Ivy's ability to grow in bright sun-

Poison Ivy

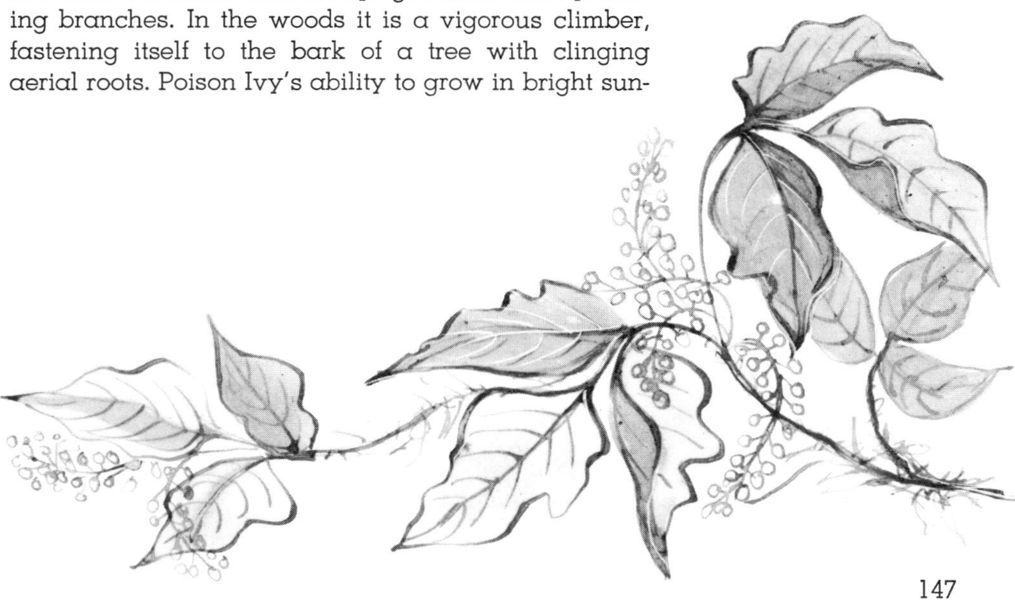

light or in shade, on dry sand or in mud, has enabled it to take over large areas of the Outer Lands.

The dune thickets shelter a varied lot of insects and birds. Tent caterpillars weave their silken canopies on the branches of the cherry trees—tents of modest dimensions, in keeping with the small scale of the trees. Eating cherry leaves, the hairy black-and-red caterpillars provide food for wasps, beetles and flies. In years when they are particularly abundant, Yellow-billed Cuckoos and Brown Thrashers fly to the dunes to feed on them.

Tent caterpillars seem to have a population explosion once every seven years. Then their tents envelop the cherry trees and they strip the branches bare. But the trees put out new leaves and the enemies of the caterpillars—chiefly parasitic wasps—keep them in control for another seven years.

Song Sparrows (Plate 13) nest in the Beach Plum bushes, closer to the sea than any other land birds. Their melodious song is punctuated on occasion by the whistled call of "Bobwhite!", as a covey of quail chicks (Plate 13) scatters through the thicket for a lesson in beetle-catching. Migrant birds visit the dunes in late summer. Then Robins and Cedar Waxwings cover the branches of the cherries and Tree Swallows and Myrtle Warblers gobble the bayberries. Flocks of Myrtle Warblers remain through the winter, eating the wax-covered fruit. They are joined by Downy Woodpeckers and Flickers who turn to the white Poison Ivy berries after insects grow scarce.

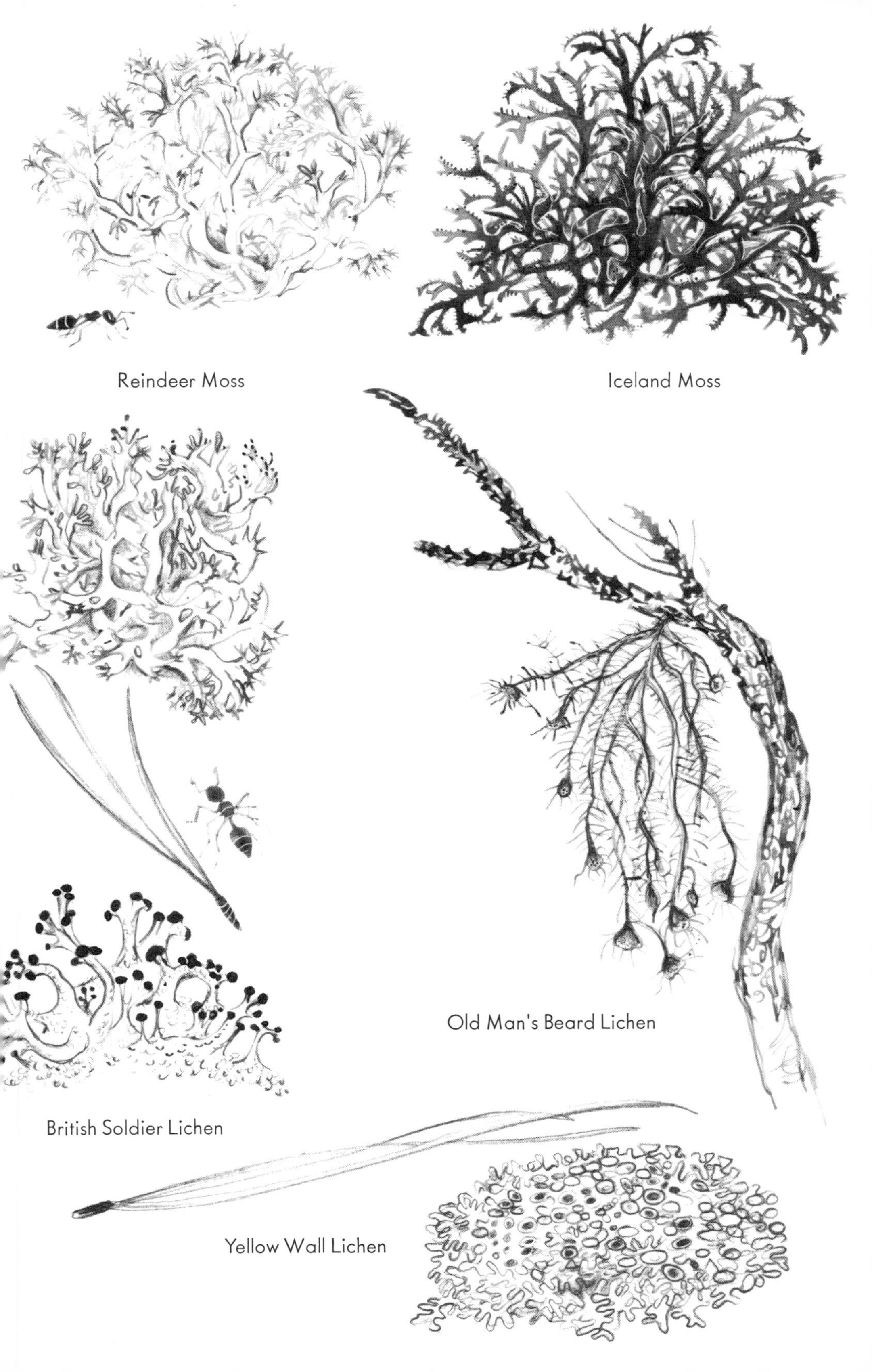
Reindeer Moss
Iceland Moss
Old Man's Beard Lichen
British Soldier Lichen
Yellow Wall Lichen

Golden Heather

Beach Heather

Heather

When a second dune rises behind the foredune, Bearberry (Plate 14), lichens and Beach Heather spread on its windward slope. Plants from the Far North, they were among the first to grow in the barren till of the Outer Lands after the glacier's retreat. Slow-growing, low, hardy, they still hold their own on the sun-baked, wind-blown sands.

Bearberry vines cross and crisscross, their gnarled woody stems weighting down the sand. The leathery evergreen leaves, which turn bronze in the fall, help the plant to retain water. When the leaves gradually drop off and decay they add a thin layer of humus to the sand. The pink Bearberry flowers produce bright red berries, sometimes called Hog Cranberries. Too mealy for human consumption, they are eaten by game birds and deer.

A clump of lichen is an intricate partnership of plants in which tangled fungus threads wrap around single-celled green algae. This close association, in which both partners benefit, is called symbiosis. The algae manufacture food while the fungi supply water and anchorage. Lacking roots, leaves and flowers, lichens spread when fragments of the colony break off and are carried away by the wind. They also reproduce by spores and by minute, powdery granules which look like dust on the surface of the plant colony.

Demanding little except light and water, lichens thrive in the moisture-laden atmosphere of the Cape and islands. Inch-high British Soldiers march along sandy paths. Tufts of Old Man's Beard hang from the

Broom Crowberry

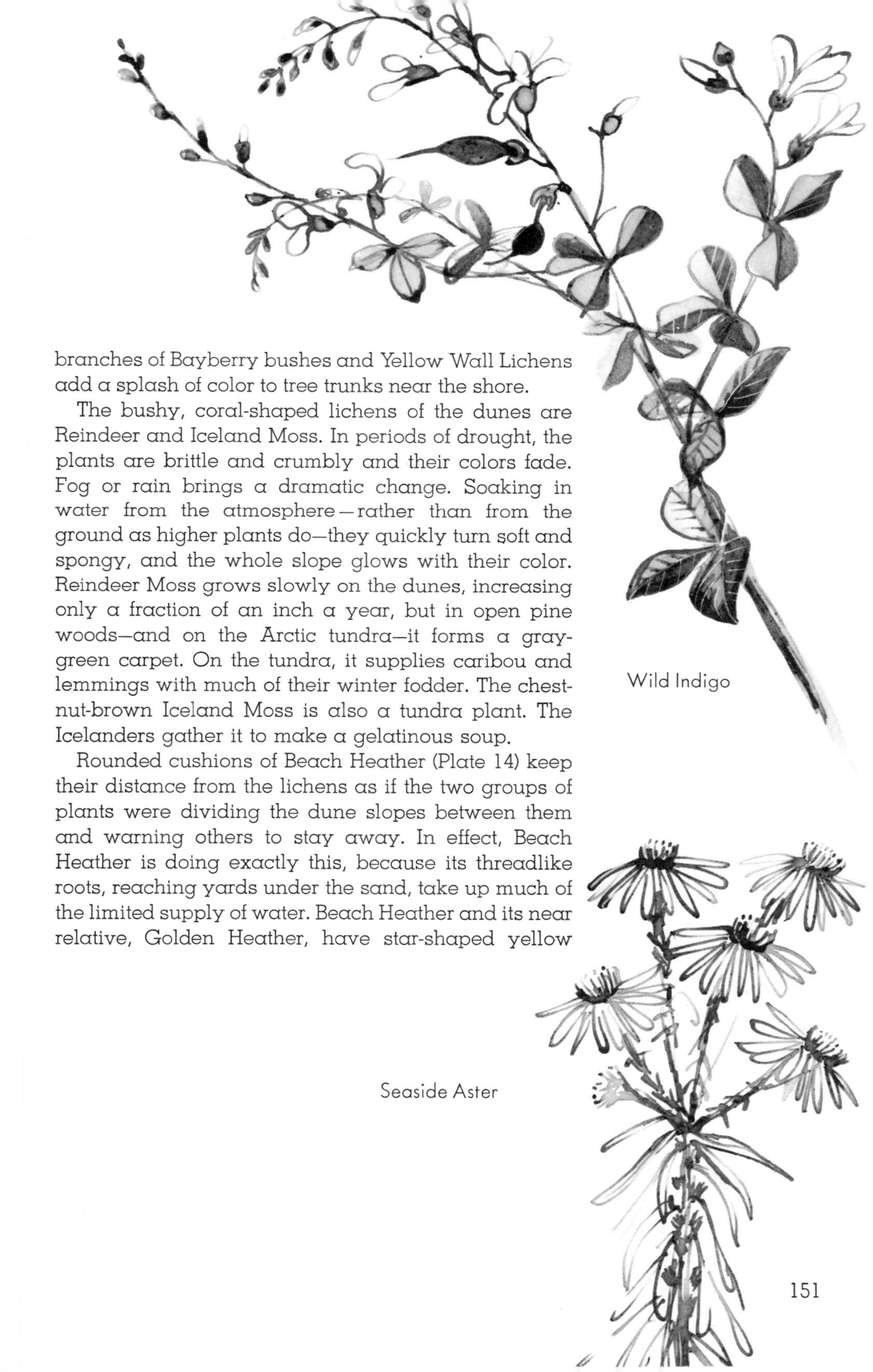

branches of Bayberry bushes and Yellow Wall Lichens add a splash of color to tree trunks near the shore.

The bushy, coral-shaped lichens of the dunes are Reindeer and Iceland Moss. In periods of drought, the plants are brittle and crumbly and their colors fade. Fog or rain brings a dramatic change. Soaking in water from the atmosphere—rather than from the ground as higher plants do—they quickly turn soft and spongy, and the whole slope glows with their color. Reindeer Moss grows slowly on the dunes, increasing only a fraction of an inch a year, but in open pine woods—and on the Arctic tundra—it forms a gray-green carpet. On the tundra, it supplies caribou and lemmings with much of their winter fodder. The chestnut-brown Iceland Moss is also a tundra plant. The Icelanders gather it to make a gelatinous soup.

Rounded cushions of Beach Heather (Plate 14) keep their distance from the lichens as if the two groups of plants were dividing the dune slopes between them and warning others to stay away. In effect, Beach Heather is doing exactly this, because its threadlike roots, reaching yards under the sand, take up much of the limited supply of water. Beach Heather and its near relative, Golden Heather, have star-shaped yellow

Wild Indigo

Seaside Aster

Earth Stars

flowers that open on sunny days in May and June. The sage-green, scalelike leaves of Beach Heather are covered with a woolly down, while Golden Heather leaves are darker green and less hairy.

Although Beach Heather is sometimes called Poverty Grass because it grows where little else could survive, it is neither a grass nor a heather. True heather comes from Europe and Asia Minor and rarely grows wild in the United States—except on Nantucket. Brought there by accident along with a shipment of pines from Scotland, it has become naturalized on the island's sandy moors. Its purple or pink bell-shaped flowers bloom all summer.

Broom Crowberry is another low-growing shrub of the dunes which is sometimes mistaken for Beach Heather. A survivor of a period even earlier than the Ice Age, it grows in only a few places along the Atlantic coast. Its purplish-brown flowers, blooming early in spring, are followed by small berries.

Between patches of lichen and Beach Heather, a handful of other flowers bloom: Golden Asters (Plate 14) whose narrow leaves and woolly stems help them fight drought, purple Seaside Asters, Bushy Wild Indigo and Prickly Pears. Prickly Pears (Plate 14) are cactuses that have migrated from the hot dry deserts of the Southwest to the shores of the Atlantic Ocean. Shedding their minute leaves, they depend on their stems—the flat spiny pads that sprawl across the sand—for food and water storage. Their butter-yellow blossoms which wither in a day are followed by juicy red "pears."

Rose Pogonia

Another unusual tenant of this dune neighborhood is the Earth Star, a small mushroom related to puffballs. As the Earth Star matures, its tough outer coat splits into triangular segments. In wet weather the segments unfold like the petals of a flower and the Earth Star clings to the sand, absorbing moisture. On dry days, its "petals" curl up again, covering the ball in the center which contains spores. When the wind blows the mushroom across the dunes, its ripe spores spill out through a hole in the top of the ball.

On warm evenings, the dune slopes are alive with toads (Plate 14). Toads migrate to ponds, or to fresh-marsh pools in late spring and early summer to lay eggs. After their tadpoles develop, regiments of toads —little ones, middle-sized ones, big ones—return to the dunes. Hiding under boards or burying themselves during the daylight hours, they hop across the sands at night. Unblinking, a toad waits for a mosquito or moth to come within range. Then its sticky tongue darts out and with one deft movement it flips the insect into its mouth. The slow-moving toads have few defenses against birds or snakes. When they cannot escape by jumping, they freeze. If turned on its back, a toad will stiffen its legs, hold its breath and play dead for minutes at a time.

Indian Pipes

Partridge Berry

Moccasin Flower

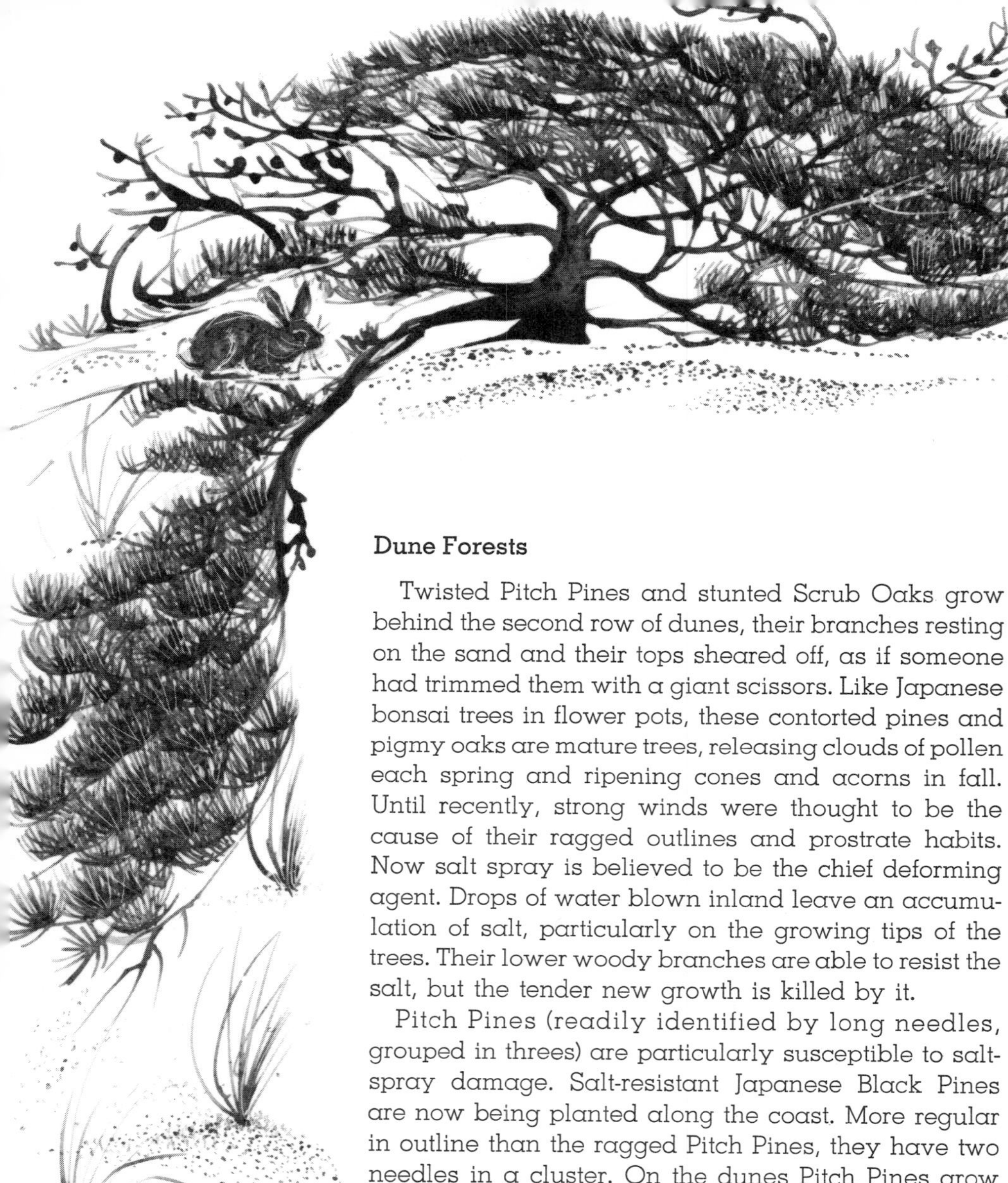

Dune Forests

Twisted Pitch Pines and stunted Scrub Oaks grow behind the second row of dunes, their branches resting on the sand and their tops sheared off, as if someone had trimmed them with a giant scissors. Like Japanese bonsai trees in flower pots, these contorted pines and pigmy oaks are mature trees, releasing clouds of pollen each spring and ripening cones and acorns in fall. Until recently, strong winds were thought to be the cause of their ragged outlines and prostrate habits. Now salt spray is believed to be the chief deforming agent. Drops of water blown inland leave an accumulation of salt, particularly on the growing tips of the trees. Their lower woody branches are able to resist the salt, but the tender new growth is killed by it.

Pitch Pines (readily identified by long needles, grouped in threes) are particularly susceptible to salt-spray damage. Salt-resistant Japanese Black Pines are now being planted along the coast. More regular in outline than the ragged Pitch Pines, they have two needles in a cluster. On the dunes Pitch Pines grow almost horizontally, becoming taller and taller as they move away from the water. Out of the range of salt spray they can be fifty feet tall. Scrub Oaks, as their name implies, never become tall trees. Even in thickets on the moors, they are no more than ten feet high.

Pitch Pine

Scrub Oak

In the lee of the dunes, the diminutive pines and oaks create their own forest community. Often a single struggling pine prepares the way. Its falling needles enrich the barren sand and give grass and flower seeds a chance to sprout. Like an oasis in the desert, a circle of green surrounds the tree. The oasis enlarges as other pine and oak seedlings root along its borders. Rabbits nibble on the grass and leave their droppings behind. When a dead tree limb topples to the ground, a termite colony tunnels through it. By hastening the process of decay, the termites do their bit to add nutrients to the soil.

As the pine and oak branches spread, the oasis becomes too shady for the sun-loving grasses and dune flowers. Then pink Moccasin Flowers and trailing Partridge Berries bloom under the dwarfed trees, and ghostly Indian Pipes, plants of the deep woods, send up their waxen stalks. Although Indian Pipes are flowering plants, they lack chlorophyll and cannot make their own food. Like mushrooms, they get nourishment from decaying vegetation in the soil.

Foxes visit the knee-high forest at night. By day, Towhees scratch among the rusty pine needles on the ground. Catbirds hunt along the twisted branches and Blue Jays fly off with pine seeds and acorns.

Although these vest-pocket woods are scattered over the dunes of the Outer Lands, none is as spectacular as the Sunken Forest on Fire Island. Less than 500 feet from the ocean a path leads abruptly from the bright glare of sand and water to a world of darkness where the air feels cool even on a summer day. Here are stout Hollies with glossy leaves and berries, Red Cedars and Sassafras, maple, oak and pine. Crowded together in a hollow which slopes away from the dunes, the trees grow up to the height of the dune in front of them, and no further. Undisturbed for centuries, they have laid down a spongy carpet of leaf mold on the sand. Vines rooting there have curtained over the woods. Catbrier, Grape, Poison Ivy wind around the trunks and loop from tree to tree, shutting out the light. Where some sunshine filters through, Blueberries and Shadbushes ripen their berries, and ferns and Canada Mayflowers spread over the shadowy forest floor. A refuge for birds, butterflies—and mosquitoes, the Sunken Forest is a reminder of what the woods of the Outer Lands were like before Europeans settled along their sandy shores.

Anchors Aweigh

Few places are as subject to sudden change as the dunes. Although dune life tends to progress from bare sand to dense woods, a hurricane, a fire or the hand of man can halt the progression and destroy centuries of growth in short order. The death of a single clump of Beach Grass on the crest of a dune may be enough to free the sand hill from its bonds and set it in motion. When a blowout—a break in the mat of vegetation—occurs, the wind quickly enlarges it. Whipping through the break, it bares the roots of neighboring plants. A second grass clump dies, and a third. As more and more sand is released from the grip of the grass roots, the whole dune begins to wander.

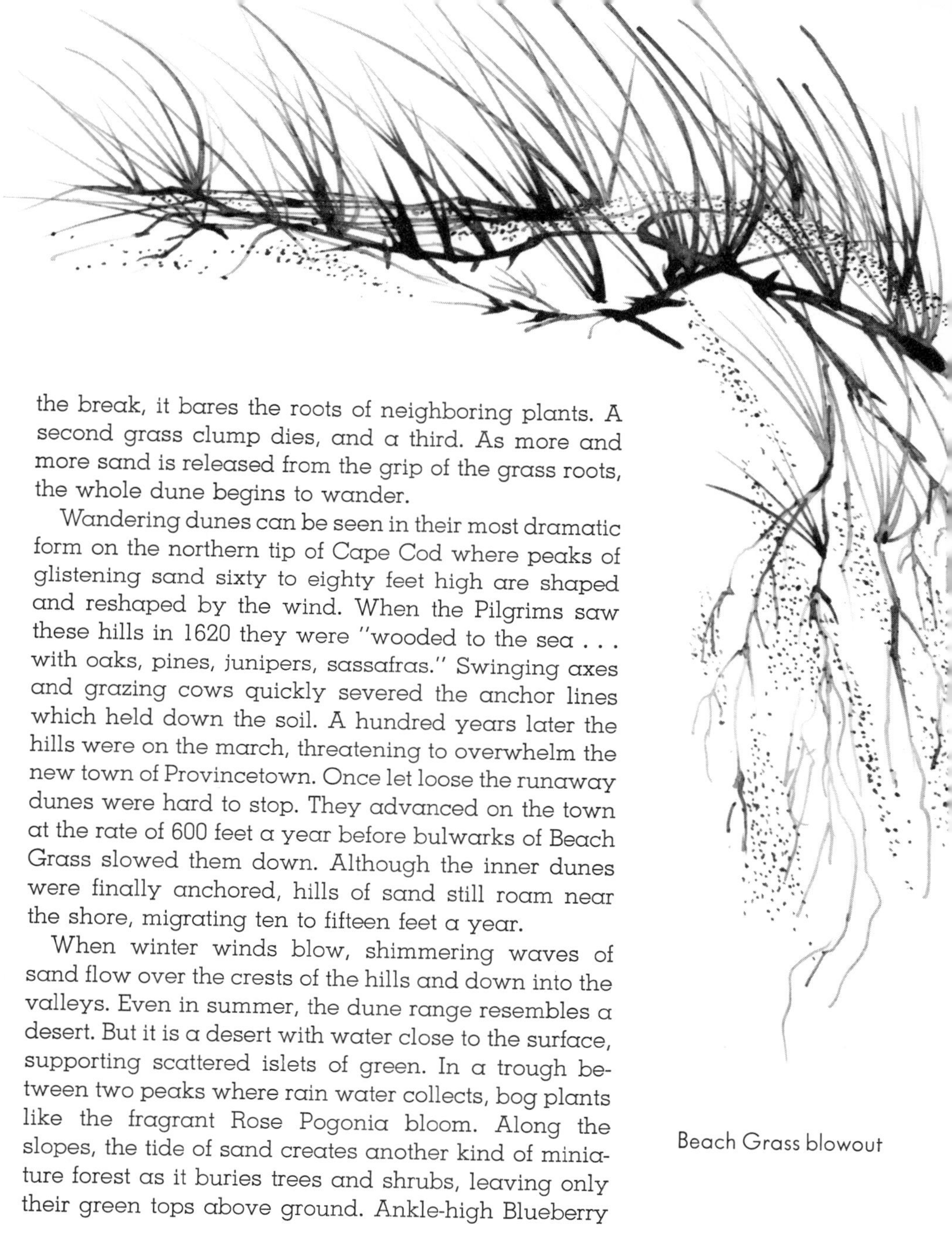

Wandering dunes can be seen in their most dramatic form on the northern tip of Cape Cod where peaks of glistening sand sixty to eighty feet high are shaped and reshaped by the wind. When the Pilgrims saw these hills in 1620 they were "wooded to the sea . . . with oaks, pines, junipers, sassafras." Swinging axes and grazing cows quickly severed the anchor lines which held down the soil. A hundred years later the hills were on the march, threatening to overwhelm the new town of Provincetown. Once let loose the runaway dunes were hard to stop. They advanced on the town at the rate of 600 feet a year before bulwarks of Beach Grass slowed them down. Although the inner dunes were finally anchored, hills of sand still roam near the shore, migrating ten to fifteen feet a year.

When winter winds blow, shimmering waves of sand flow over the crests of the hills and down into the valleys. Even in summer, the dune range resembles a desert. But it is a desert with water close to the surface, supporting scattered islets of green. In a trough between two peaks where rain water collects, bog plants like the fragrant Rose Pogonia bloom. Along the slopes, the tide of sand creates another kind of miniature forest as it buries trees and shrubs, leaving only their green tops above ground. Ankle-high Blueberry

Beach Grass blowout

bushes grow like vines, with their fruit resting on the sand, and grasses tower above Cherries and Pines. The forest is short-lived, for the trees cannot keep pace with the sand that engulfs them. Years later as the dune moves on, the buried forest is uncovered again—a graveyard now with skeleton branches and trunks as silvery as the sand around them.

Buried forest, Provincetown

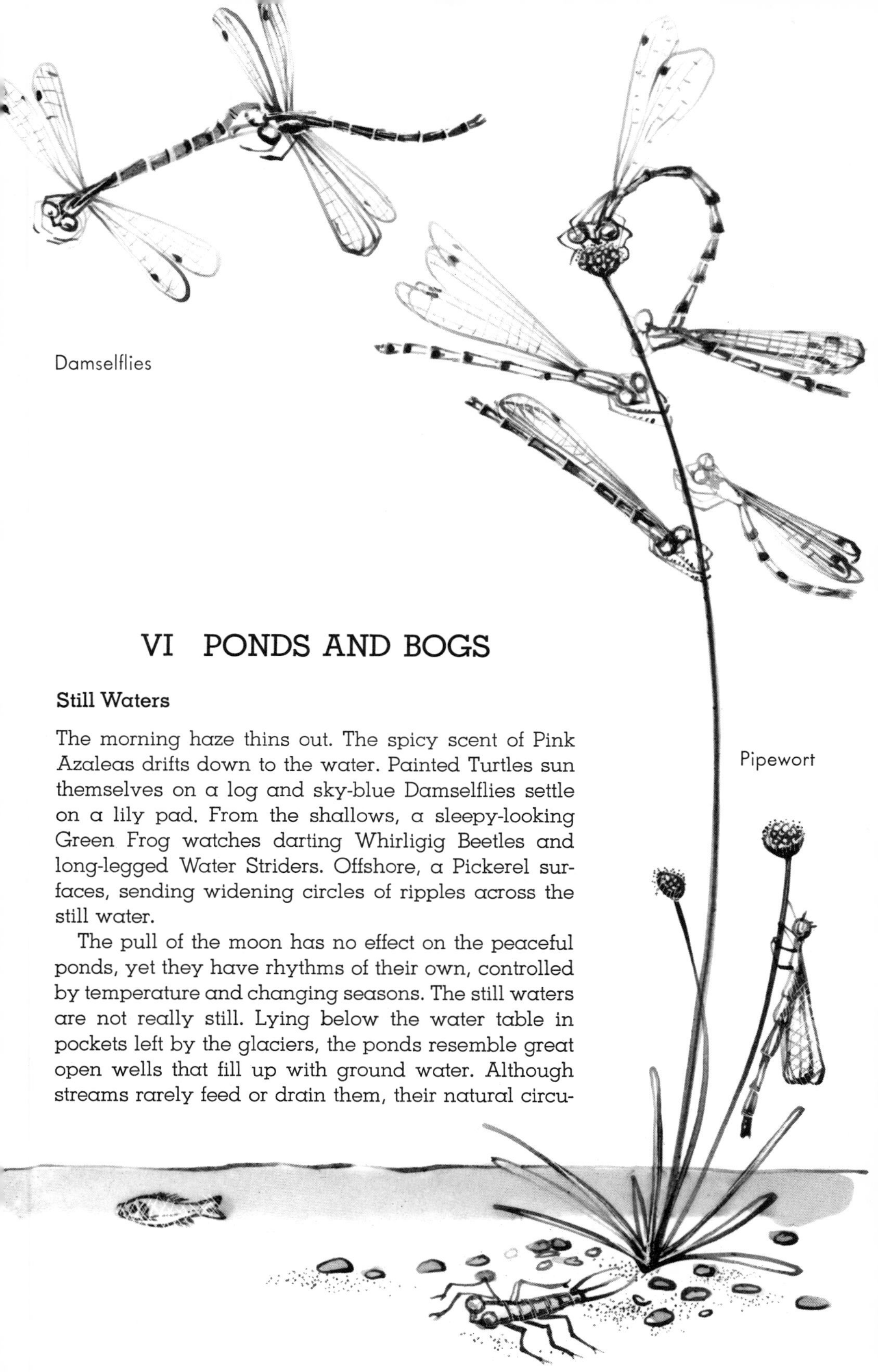

Damselflies

Pipewort

VI PONDS AND BOGS

Still Waters

The morning haze thins out. The spicy scent of Pink Azaleas drifts down to the water. Painted Turtles sun themselves on a log and sky-blue Damselflies settle on a lily pad. From the shallows, a sleepy-looking Green Frog watches darting Whirligig Beetles and long-legged Water Striders. Offshore, a Pickerel surfaces, sending widening circles of ripples across the still water.

The pull of the moon has no effect on the peaceful ponds, yet they have rhythms of their own, controlled by temperature and changing seasons. The still waters are not really still. Lying below the water table in pockets left by the glaciers, the ponds resemble great open wells that fill up with ground water. Although streams rarely feed or drain them, their natural circu-

lation keeps them from becoming stagnant. In part this is due to the nature of water itself, which is heavier when it is cold. (Water is heaviest at a temperature of 39.2 degrees. Below that temperature it becomes lighter and less dense.) After sunset on summer nights, the surface water cools off. Dropping down, it forces the warmer water below to rise. This turnover becomes faster in spring and fall when temperature changes are greater. Tiny plants and animals travel with the water, moving up in spring and down in fall as if they were riding on an elevator.

In addition to this slow but regular circulation, the amount of oxygen in the water varies throughout the day and year, and from the top to the bottom of the pond. Winds and waves, acting like giant eggbeaters, aerate the water of the bays, but a pond depends largely on green plants to maintain its oxygen supply. As the plants manufacture food, the energy they get from the sun enables them to split molecules of water into hydrogen and oxygen. Retaining the hydrogen, they release quantities of oxygen into the pond.

The tide of life rises and falls with the increase and decrease of the pond's oxygen supply. It is high in spring when a thousand thousand creatures, cold-blooded and warm, finned, winged, many-legged, awaken from a winter's hibernation to mate and lay eggs. It rises to the full in summer as eggs hatch and flowers bloom, as tadpoles become frogs and Midges and Mayflies swarm. With the coming of fall, the bass

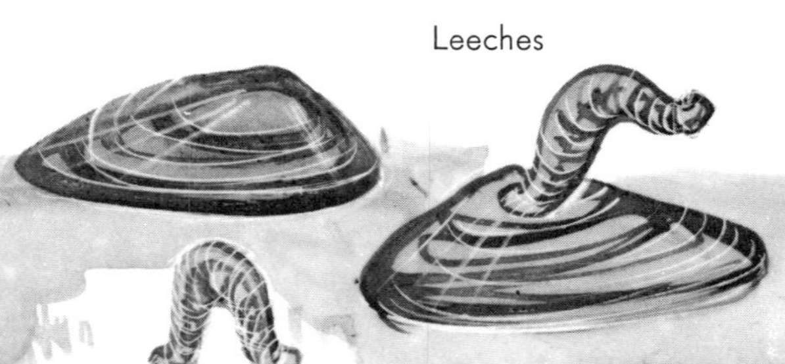

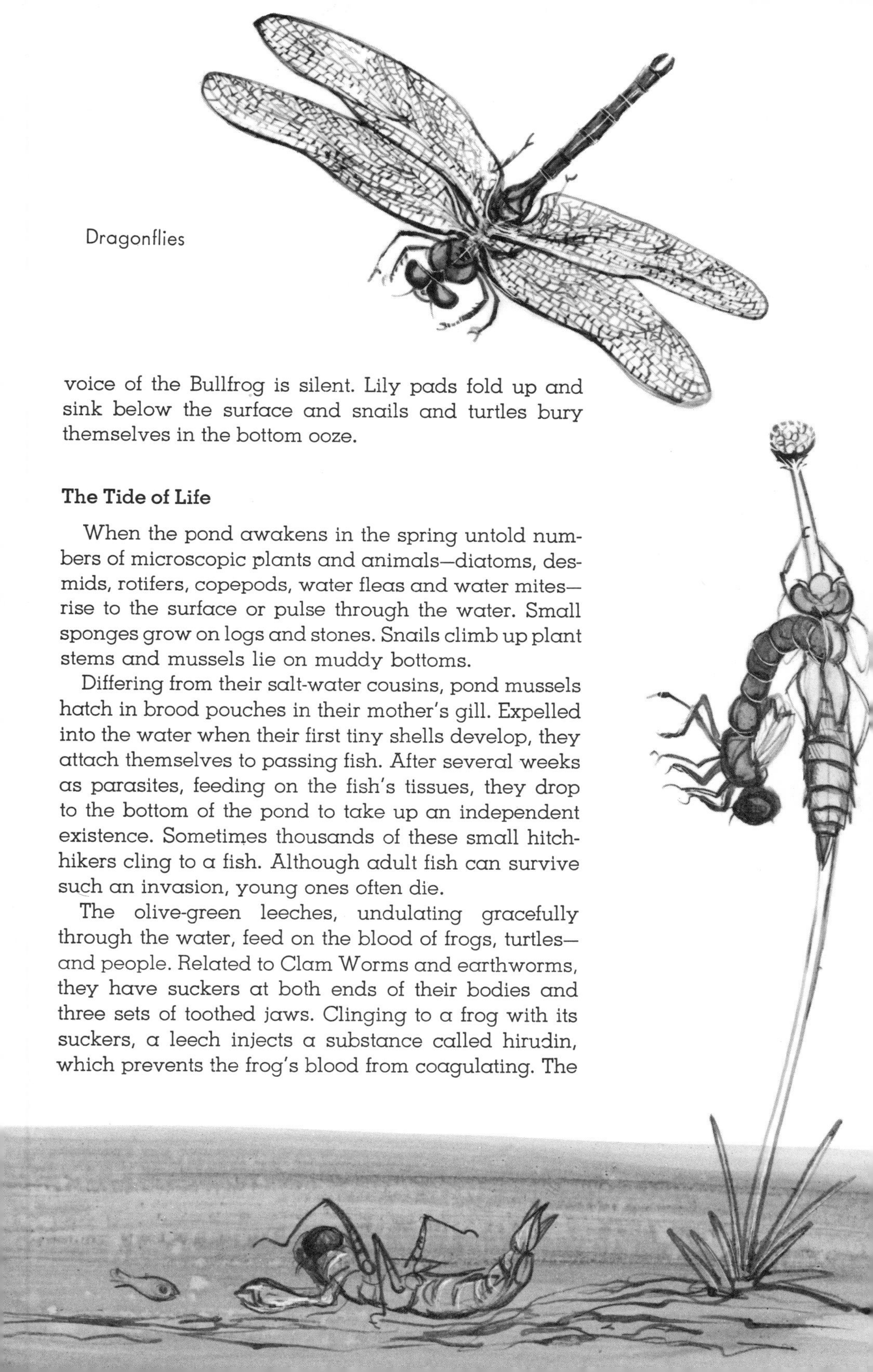

Dragonflies

voice of the Bullfrog is silent. Lily pads fold up and sink below the surface and snails and turtles bury themselves in the bottom ooze.

The Tide of Life

When the pond awakens in the spring untold numbers of microscopic plants and animals—diatoms, desmids, rotifers, copepods, water fleas and water mites—rise to the surface or pulse through the water. Small sponges grow on logs and stones. Snails climb up plant stems and mussels lie on muddy bottoms.

Differing from their salt-water cousins, pond mussels hatch in brood pouches in their mother's gill. Expelled into the water when their first tiny shells develop, they attach themselves to passing fish. After several weeks as parasites, feeding on the fish's tissues, they drop to the bottom of the pond to take up an independent existence. Sometimes thousands of these small hitchhikers cling to a fish. Although adult fish can survive such an invasion, young ones often die.

The olive-green leeches, undulating gracefully through the water, feed on the blood of frogs, turtles—and people. Related to Clam Worms and earthworms, they have suckers at both ends of their bodies and three sets of toothed jaws. Clinging to a frog with its suckers, a leech injects a substance called hirudin, which prevents the frog's blood from coagulating. The

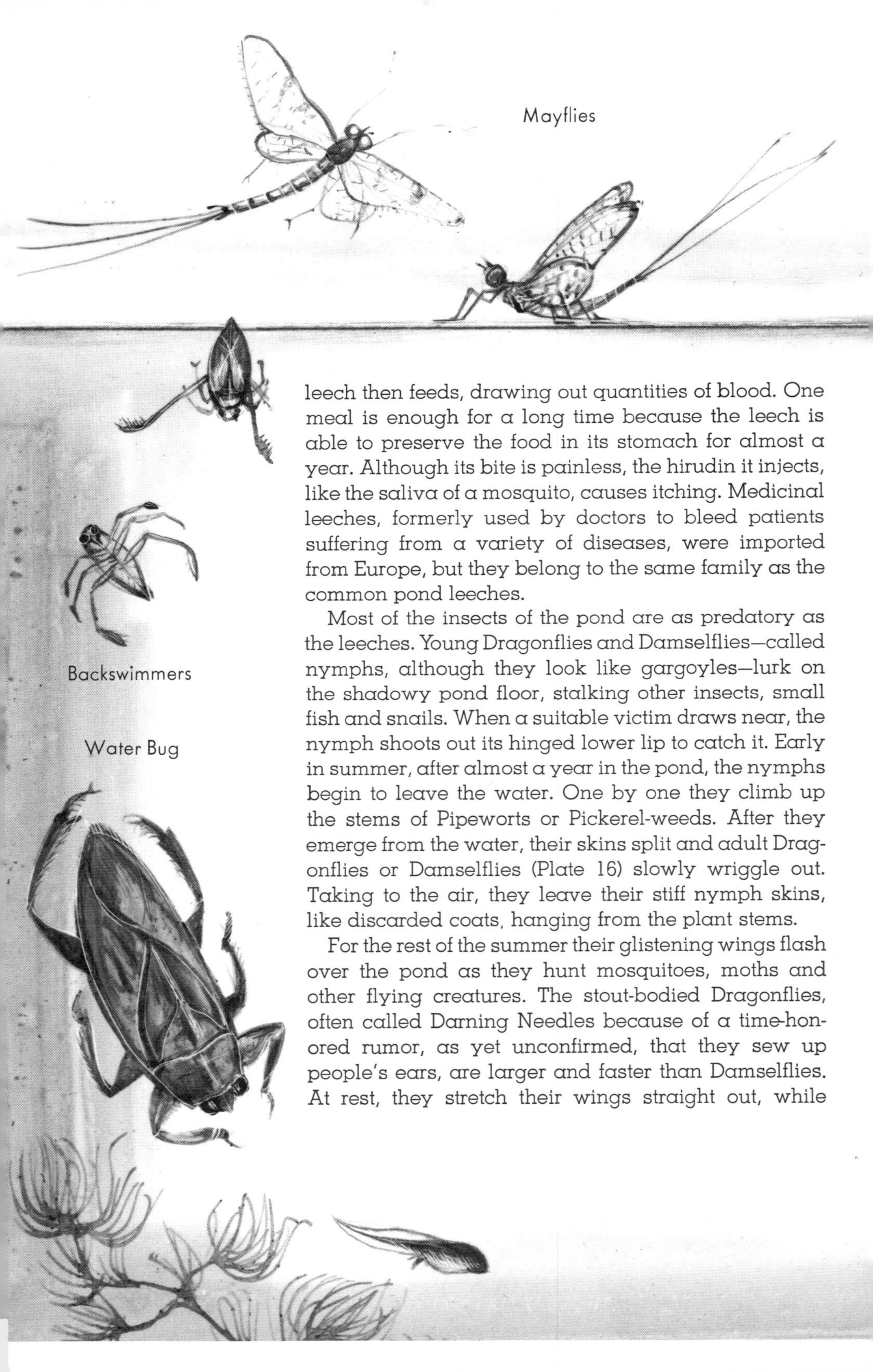

leech then feeds, drawing out quantities of blood. One meal is enough for a long time because the leech is able to preserve the food in its stomach for almost a year. Although its bite is painless, the hirudin it injects, like the saliva of a mosquito, causes itching. Medicinal leeches, formerly used by doctors to bleed patients suffering from a variety of diseases, were imported from Europe, but they belong to the same family as the common pond leeches.

Most of the insects of the pond are as predatory as the leeches. Young Dragonflies and Damselflies—called nymphs, although they look like gargoyles—lurk on the shadowy pond floor, stalking other insects, small fish and snails. When a suitable victim draws near, the nymph shoots out its hinged lower lip to catch it. Early in summer, after almost a year in the pond, the nymphs begin to leave the water. One by one they climb up the stems of Pipeworts or Pickerel-weeds. After they emerge from the water, their skins split and adult Dragonflies or Damselflies (Plate 16) slowly wriggle out. Taking to the air, they leave their stiff nymph skins, like discarded coats, hanging from the plant stems.

For the rest of the summer their glistening wings flash over the pond as they hunt mosquitoes, moths and other flying creatures. The stout-bodied Dragonflies, often called Darning Needles because of a time-honored rumor, as yet unconfirmed, that they sew up people's ears, are larger and faster than Damselflies. At rest, they stretch their wings straight out, while

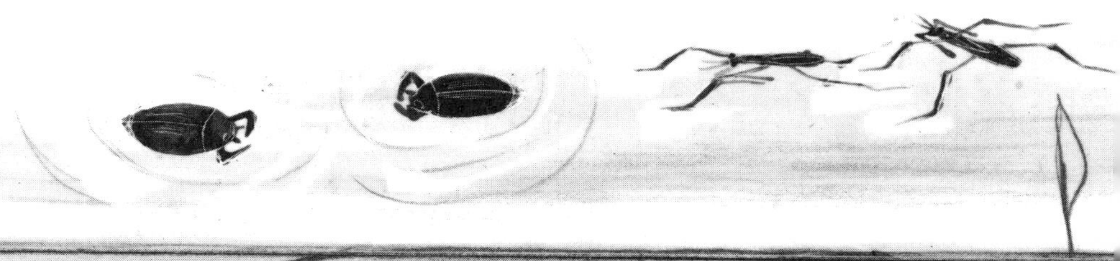

most Damselflies fold theirs over their backs. Damselflies fly in tandem after mating, the male escorting the female when she inserts her eggs in the stems of water plants, often ducking underwater to find a safe place. Some Dragonfly females are also escorted by mates; others fly alone, skimming over the surface of the pond and dropping their eggs in the water.

The surface insects—the Water Striders who skate across the pond, their feet dimpling the elastic water skin, circling Whirligigs and rowing Backswimmers, big Water Bugs and Diving Beetles—are also hunters, attacking not only insects but tadpoles and salamanders as well. They are all air-breathers who carry silvery bubbles of air with them when they swim underwater. Strong fliers, they migrate to other ponds late in summer if the one they have been living in becomes overcrowded.

Frail Mayflies, on the other hand, are defenseless. As nymphs they live on the pond bottom, feeding largely on plankton plants and sprinting to escape from Dragonflies, turtles, fish and even wading birds. Those who survive leave the pond at the same time, often in awesome numbers. After a final molt, they take part in a spectacular mating flight. The transparent insects rise and fall, rise and fall, with their long tails trailing, while birds and bats weave in and out of their swarm and fish wait below, ready to snap up any of the dancing insects that fall to the water. Unequipped for eating, Mayflies live for only a day or two after mating.

Diving Beetle

Pondweed

Snapping Turtle

Painted Turtle

Spotted Turtle

The females usually float on the surface of the pond with wings outstretched when they lay eggs.

From early spring to mid-summer, the mating calls of frogs and toads reverberate from the water and the nearby woods as males summon females from their winter resting places. Trilling Spring Peepers start the nightly chorus and the Bullfrog's "jug-o-rum" and "be drowned, better go round" provide the final stanzas. After mating, the females lay round jelly-covered eggs in the water. While the eggs hatch and the fishlike tadpoles develop, most of their parents return to the woods and fields. Green Frogs (Plate 16) (who are often more brown than green) and Bullfrogs remain in the pond, floating with eyes and nostrils above the surface, or hiding in the shallows where insects and snails abound. The two pond frogs resemble each other, but the smaller Green Frog has two ridges of folded skin along its back, while the Bullfrog's back is smooth.

At the same time that frogs are calling, fish spawn in the shallows and turtles lay their eggs on shore. Yellow Perch lay their eggs in long jelly-covered strings, twisted around waterweeds, but Sunfish and Bass prepare nests first. The brightly colored male Sunfish clears a space on the bottom of the pond, sweeping away pebbles with his tail and using his mouth to move larger stones. He remains to guard the eggs after they are laid, while the female swims away.

Although the turtles of the pond—and some woods turtles—mate in the water, they travel to shore for egg-laying. Scooping out holes with their hind legs, females bury their eggs in the soil, where the sun warms them. As soon as they break out of their shells, the inch-long hatchlings head for the water. Painted Turtles and the polka-dotted Spotted Turtles are pleasant pondside companions when they climb up on logs to bask in the sun, or swim near the surface, feeding on plants and small animals. Snapping Turtles—which may weigh fifty pounds when full-grown—patrol the bottom of the pond, striking out with their hooked beaks at fish and frogs and pulling ducklings underwater. Although they rarely snap if they are stepped on when they are in the pond, they are capable of formidable action on land where they can rear up and lunge forward with startling speed.

Bullfrog

Belted Kingfisher

Birds

Black Ducks and Mallards breed on the marshy edges of the ponds. Belted Kingfishers nest in clay or gravel banks overhanging the water, digging deep tunnels to shelter their young. Most of the birds around the ponds, however, are visitors from other neighborhoods—Pewees, Kingbirds and warblers from the woods, Swifts and Barn Swallows (Plate 16) from the towns, who come to feed on the swarms of insects that fly up from the water. Herring Gulls make the ponds a point of call on their daily trips between beach and garbage dump and at night Whippoorwills whistle and owls hunt frogs. Late in summer the bird population builds up as Loons, Canada Geese and Ring-necked Ducks arrive from the north, to remain until the ponds freeze over in winter.

Pond Plants

Woody shrubs such as Pink Azalea (Plate 15) and Highbush Blueberry, Sweet Pepper Bush, Buttonbush (Plate 15) and Sheep Laurel (Plate 15) border the ponds, and spikes of Meadowsweet (Plate 15) and Hardhack bloom in low wet places. But the other pond plants are all water-dwellers. Unlike the simple seaweeds, they are flowering plants whose ancestors once lived on land but who, during a long evolution, have become aquatic. Some wade in the shallows while others float freely without even holdfasts to anchor them.

Pickerel-weeds (Plate 16), Pipeworts and Arrowheads (Plate 16) live only a step away from land. Retaining the rigid skeletons of land plants, they hold their leaves and flowers above water. Amphibians like the frogs, they can live in wet soil as well as in water.

Pondweeds, rooting in the bottom mud, live almost entirely under the water. Although they send inconspicuous flower stalks up into the air where the wind pollinates them, their submerged leaves respire as fish do, by using oxygen dissolved in the water. Their long slender leaves provide shelter for fish and food for waterfowl and snails. Waterweeds—the familiar plants of aquariums—are true aquatics. Loosely rooted, they drift through the deeper water. When their brittle stems break, each fragment forms a new plant. Vigorous growers, Waterweeds often crowd out other vegetation and make swimming and rowing difficult.

Duckweeds, whose family includes the smallest of all flowering plants, float on the surface with their roots dangling. *Wolffia*, the smallest Duckweed, is a speck of green, one-twenty-fifth of an inch in diameter. Four and a half million *Wolffias* weigh a pound. Although the Duckweeds have tiny flowers, they usually reproduce by budding—their flat plant bodies dividing into two parts like an amoeba. This division takes place so rapidly that by mid-summer astronomical numbers of Duckweeds form floating mats of green. Round bulblets, developing on the Duckweeds in fall, sink to the

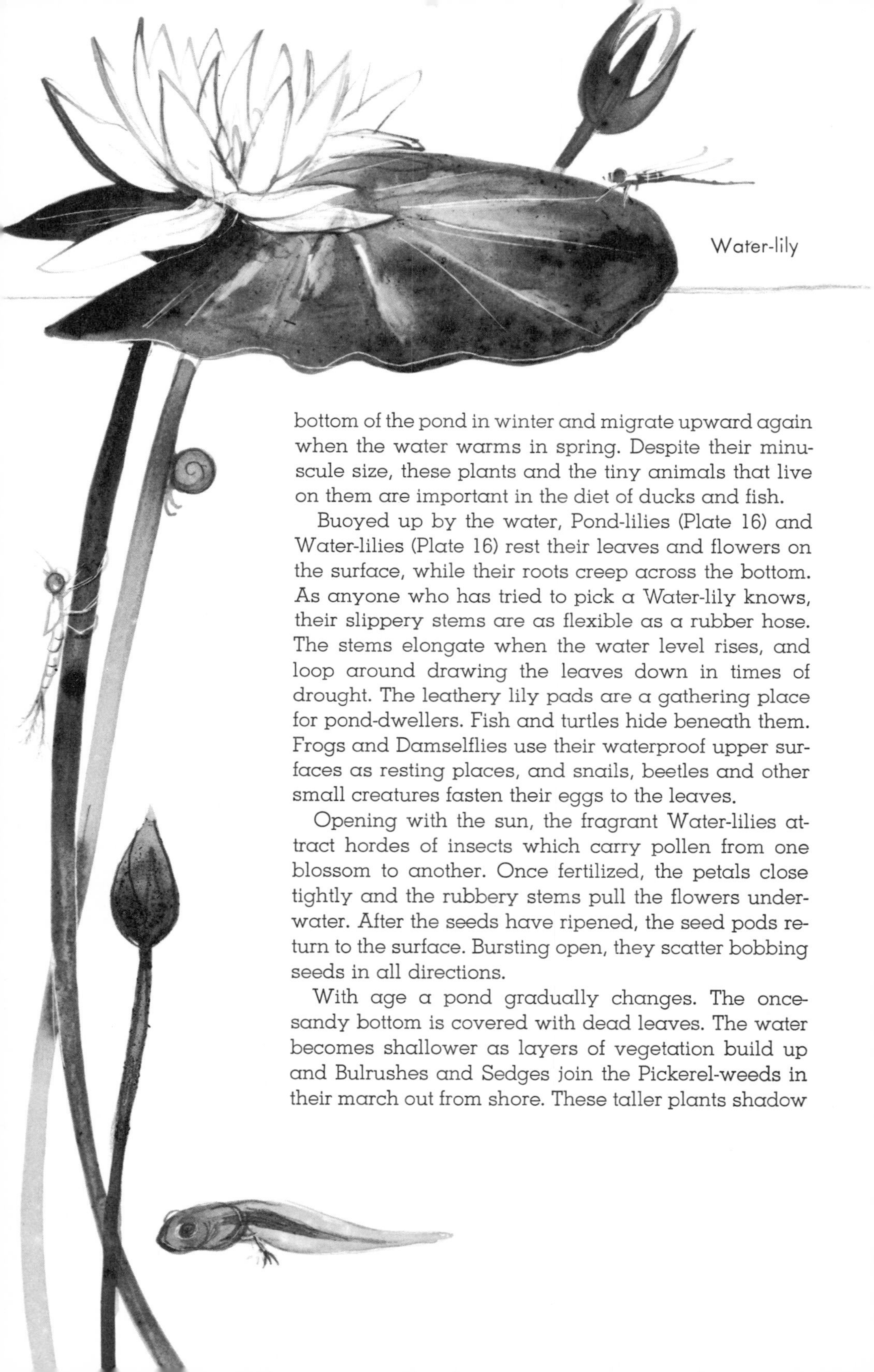

Water-lily

bottom of the pond in winter and migrate upward again when the water warms in spring. Despite their minuscule size, these plants and the tiny animals that live on them are important in the diet of ducks and fish.

Buoyed up by the water, Pond-lilies (Plate 16) and Water-lilies (Plate 16) rest their leaves and flowers on the surface, while their roots creep across the bottom. As anyone who has tried to pick a Water-lily knows, their slippery stems are as flexible as a rubber hose. The stems elongate when the water level rises, and loop around drawing the leaves down in times of drought. The leathery lily pads are a gathering place for pond-dwellers. Fish and turtles hide beneath them. Frogs and Damselflies use their waterproof upper surfaces as resting places, and snails, beetles and other small creatures fasten their eggs to the leaves.

Opening with the sun, the fragrant Water-lilies attract hordes of insects which carry pollen from one blossom to another. Once fertilized, the petals close tightly and the rubbery stems pull the flowers underwater. After the seeds have ripened, the seed pods return to the surface. Bursting open, they scatter bobbing seeds in all directions.

With age a pond gradually changes. The once-sandy bottom is covered with dead leaves. The water becomes shallower as layers of vegetation build up and Bulrushes and Sedges join the Pickerel-weeds in their march out from shore. These taller plants shadow the Water-lilies and Pondweeds until they no longer receive enough sunlight to make food. More leaves drop to the pond floor, more wading plants move in, and the water becomes still shallower. Over the centuries, many small ponds, particularly on Nantucket and Block Island, have been transformed into marshes or bogs.

Bogs

Bogs develop when clumps of Sphagnum Moss, accompanying the Sedges and Rushes, form floating mats that roof over the ponds. Soaking up extraordinary amounts of water—two hundred times their own weight—the mosses grow at the surface and die below. Cut off from oxygen by the spongy mat above them, their dead leaves and stems form peat rather than soil. Peat was the principal fuel in much of the Outer Lands until coal stoves were introduced. Block Islanders called it "tug," referring to the hard work entailed in getting it out of the bog. Although bog peat is rich in carbon—which is why it is valuable as fuel—it lacks other chemicals necessary for plant growth. Only certain plants can survive in the acid peat bogs.

Cranberry vines root in the Sphagnum, their woody stems trailing over the squashy moss. Their flower buds, resembling the beak and head of a crane, open in early summer. In fall when their tart berries ripen,

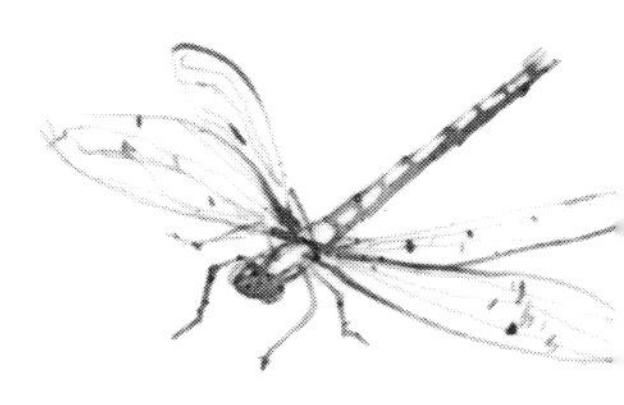

Damselfly

Duckweed

Sunfish

their leaves turn wine-red. Wild Cranberries grow in bogs all over the Outer Lands, and as far west as Wisconsin, but the berries in the market are cultivated.

Early in the last century a Cape Codder noticed that cranberries grew larger and juicier if sand from the dunes blew over them. His discovery, coming at a time when sailing vessels were being replaced by steamboats and the Cape's economy was shaky, touched off a small-scale gold rush. Prices of bog lands soared as beached sea-captains and their crews turned to cranberry farming. Mucky acres were cleared and ditched* and schools opened late in the fall so that even children could turn out for the harvest. Today, although mechanical harvesters are replacing old-fashioned hand scoops and the center of cranberry production has shifted to the mainland, Nantucket still has one of the largest cranberry bogs in the world and there are working bogs on Cape Cod and in some parts of Martha's Vineyard.

Pitcher Plants and Sundews trap and digest insects, obtaining from them the nitrogen that they cannot get

*The ditches permit growers to flood the bogs from time to time, to protect the plants against frost.

Cranberries

Sphagnum Moss

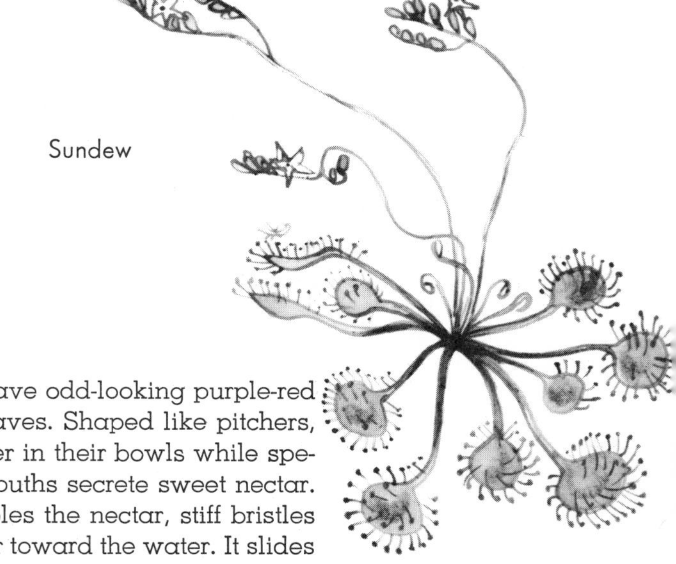

Sundew

in the bog. Pitcher Plants have odd-looking purple-red flowers and even odder leaves. Shaped like pitchers, the leaves collect rain water in their bowls while special cells at their pitcher-mouths secrete sweet nectar. When a fly or beetle samples the nectar, stiff bristles on the leaf propel the visitor toward the water. It slides down easily enough, but the curving bristles prevent it from climbing up again or finding a solid footing so that it can fly. After considerable buzzing and wing-flapping, the insect drowns and the plant slowly absorbs its remains.

The little Sundews have white flowers and glistening, hairy leaves. The dewlike drops on their leaves are a sticky substance that the hairs secrete. Any Midge landing on the leaf finds its feet caught in this mucilage. The more it tries to free itself, the more mucilage the plant pours out. Neighboring hairs also bend toward the struggling insect, gluing it to the leaf until it dies of suffocation. Sundews are sometimes found on sandy pond beaches as well as in bogs.

Decade after decade as moss grows on top of moss, soaking up water, the bog begins to dry out. Acid-tolerant shrubs take root, adding their woody remains to the beds of peat. They are followed by swamp trees

Pitcher Plant

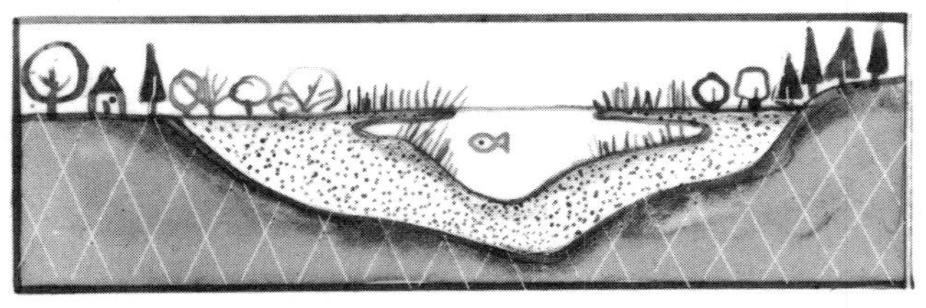

From pond to bog

—Red Maples, White Cedars, Black Gums—and eventually by hardwood forests.

The series of changes from pond to bog to dry land can be seen in all its stages on Nantucket. Most of the island is too sandy—and too windy—for deep woods, but in ancient pond beds where the soil has been enriched by decaying plants, large trees grow. Protected from northeasters by low hills, the trees, like those in the dune forests, reach up to the hilltops and no further. In the Hidden Forest, the most extensive of these tree asylums, there are still patches of Sphagnum and pools of water. But almost all of a glacial kettle hole has been taken over by gray-barked Beeches, Dogwoods and Holly, with ferns and woods flowers growing in their shade.

Given enough time, forests could conquer all of the ponds—and the rising waters of the ocean could inundate the beaches and dunes, erasing forever the work of the mighty ice sheet. But that is in a comfortably distant future. Meanwhile the Outer Lands are ours to enjoy.

APPENDIX

CHECKLIST OF SCIENTIFIC NAMES

The following are the scientific names of species described and illustrated in this book.

Sponges:

Boring Sponge	Cliona celata
Clathria Sponge	Clathria delicata
Crumb of Bread Sponge	Halichondria panicea
Eyed Finger Sponge	Chalina oculata
Mermaids' Gloves	Desmacidon palmata
Redbeard Sponge	Microciona prolifera

Coelenterates:

Hydroid on hermit crab shell	Hydractinia echinata
Moon Jelly	Aurelia aurita
Pink Jellyfish	Cyanea capillata
Plumose Anemone	Metridium dianthus
Portuguese Man-of-War	Physalia pelagica
Sagartia Anemone	Sagartia modesta
Sea anemone on hermit crab shell	Adamsia sociabilis
Sea Nettle	Dactylometra quinquecirrha
Stalked Jellyfish	Haliclystus auricula
Star Coral	Astrangia danae

Bryozoa:

Sea Lace	Membranipora pilosa
Erect Moss Animal	Bugula turrita

Mollusks:

Angel Wing	Pholas costata
Ark Shell	Arca campechiensis pexata
Bay Scallop	Pecten irradians
Boat Snail	Crepidula fornicata
Channeled Whelk	Busycon canaliculatum

Chestnut Clam	Astarte castanea
Chiton	Chaetopleura apiculata
Edible Mussel	Mytilus edulis
Edible Periwinkle	Littorina litorea
Fallen Angel	Barnea truncata
False Angel Wing	Petricola pholadiformis
Gem Clam	Gemma gemma
Hard-shell Clam	Venus mercenaria
Jingle Shell	Anomia simplex
Knobbed Whelk	Busycon caricum
Ladder Shell	Epitonium groenlandicum
Limpet	Acmaea testudinalis
Marsh Snail	Melampus bidentatus
Moon Snail	Polinices heros
Mud Snail	Nassa obsoleta
Oyster	Ostrea virginica
Oyster Drill	Urosalpinx cinerea
Pale Periwinkle	Littorina palliata
Pandora Shell	Pandora gouldiana
Pond Mussel	Anodonta cataracta
Pond Snail	Planorbis trivolvis
Purple Snail	Thais lapillus
Quahog	Venus mercenaria
Razor Clam	Ensis directus
Ribbed Mussel	Modiolus demissus plicatulus
Rough Periwinkle	Littorina rudis
Sea Scallop	Pecten magellanicus
Shipworm	Teredo navalis
Soft-shell Clam	Mya arenaria
Squid	Loligo pealli
Surf or Sea Clam	Spisula solidissima
Tellin Shell	Tellina tenera
Waved Whelk	Buccinum undatum

Annelids:

Blood Worm	Polycirrus eximius
Clam Worm	Nereis virens
Fringed Worm	Cirratulus cirratus
Hydroides dianthus	Hydroides dianthus
Leech	Macrobdella decora
Lugworm	Arenicola marina
Parchment Worm	Chaetopterus pergamentaceus
Plumed Worm	Diapatra cuprea
Spirorbis borealis	Spirorbis borealis
Trumpet Worm	Cistenides gouldii

Arthropods:

1. Crustaceans:

Acorn Barnacle	Balanus balonoides
American Lobster	Homarus amercanus
Blue Crab	Callinectus sapidus
Fiddler Crabs	Uca pugnax, pugilator and minax
Gooseneck Barnacle	Lepas fascicularis
Green Crab	Carcinides maenas
Hermit Crabs	Pagurus pollicarus and longicarpus
Jonah Crab	Cancer borealis
Lady Crab	Ovalipes ocellatus
Mole Crab	Hippa talpoida
Mud Crab	Panopeus herbstii
Oyster Crab	Pinnotheres ostreum
Pea Crab	Pinnixa chaetopterana
Rock Crab	Cancer irroratus
Sand Hoppers	Talorchestia longicornis and Orchestia agilis
Spider Crab	Libinia emarginata

2. Arachnids:

Horseshoe Crab	Limulus polyphemus
Wolf Spider	Lycosa pikei

3. Insects

Ant-lion	Myrmeleon immaculatus
Backswimmer	Notonecta undulata
Damselfly	Enallagma ebrium
Diving Beetle	Dytiscus marginalis
Dragonfly	Anax junius
Greenhead Fly	Tabanus nigrovittatus
Mayfly	Ephemera, several species
Salt-marsh Caterpillar	Estigmene acrea
Salt-marsh Mosquito	Aedes solicitans
Seaside Grasshopper	Trimerotropsis maritima
Tent Caterpillar	Malacosoma americanum
Water Bug	Lethocerus americanus
Water Strider	Gerris marginatus
Whirligig Beetle	Dineutes americanus

Echinoderms:

Blood Sea Star	Henricia sanguinolenta
Brittle Star	Ophiopholis aculeata
Common Starfish	Asterias forbesi

Green Sea Urchin	Strongylocentrotus droehbachiensis
Keyhole Urchin	Mellita testudinata
Purple Sea Urchin	Arbacia punctulata
Purple Star	Asterias vulgaris
Sand Dollar	Echinarachnius parma

Chordates:

1. The Tunicates:

Sea Grape	Molgula manhattensis
Sea Pork	Amaroucium stellatum
Sea Squirt	Ciona intestinalis

2. Vertebrates:

Fish:

Alewife	Alosa pseudoharengus
Eel	Anguilla rostrata
Pipefish	Syngathus fuscus
Sand Eel	Ammodytes americanus
Sea Horse	Hippocampus hudsonia
Skates	Raja, several species
Sunfish	Lepomis gibbosus
Torpedo Ray	Torpedo nobliana
Yellow Perch	Perca flavescens

Amphibians:

Bullfrog	Rana catesbiana
Green Frog	Rana clamitans melanota
Toads	Bufo americanus and fowleri

Reptiles:

Diamondback Terrapin	Malaclemys terrapin terrapin
Painted Turtle	Chrysemys picta
Snapping Turtle	Chelydra serpentina
Spotted Turtle	Clemmys guttata

Birds:

Bank Swallow	Riparia riparia
Barn Swallow	Hirundo rustica
Belted Kingfisher	Megaceryle alcyon
Black-backed Gull	Larus marinus
Black-crowned Night Heron	Nycticorax nycticorax

Black Duck	Anas rubripes
Black Skimmer	Rynchops nigra
Bobwhite	Colinus virginianus
Brant	Branta bernicla
Canada Goose	Branta canadensis
Clapper Rail	Rallus longirostris
Common Tern	Sterna hirundo
Flicker	Colaptes auratus
Great Blue Heron	Ardea herodias
Great Horned Owl	Bubo virginianus
Greater Yellowlegs	Totanus melanoleucus
Herring Gull	Larus argentatus
Kingbird	Tyrannus tyrannus
Laughing Gull	Larus atricilla
Long-billed Marsh Wren	Telmatodytes palustris
Loon	Gavia stellata
Mallard	Anas platyrhynchos
Marsh Hawk	Circus cyaneus
Myrtle Warbler	Dendroica coronata
Pectoral Sandpiper	Erolia melanotos
Pewee	Myiochanes virens
Piping Plover	Charadrius melodus
Red-tailed Hawk	Buteo borealis
Red-winged Blackbird	Agelaius phoeniceus
Ring-necked Duck	Nyroca collaris
Ruddy Turnstone	Arenaria interpres
Sanderling	Crocethia alba
Seaside Sparrow	Ammospiza maritima
Semi-palmated Plover	Charadrius semipalmatus
Semi-palmated Sandpiper	Ereunetes pusillus
Sharp-tailed Sparrow	Ammospiza caudacuta
Song Sparrow	Melospiza melodia
Sparrow Hawk	Falco sparverius
Spotted Sandpiper	Actitis macularia
Swift	Chaetura pelagica
Tree Swallow	Iridoprocne bicolor
Whimbrel	Numenius phaeopus
Whippoorwill	Antrostomus vociferus

Mammals:

Meadow mouse	Microtus pennsylvanicus
Mole	Scalopus aquaticus
Muskrat	Ondatra zibethica
Rabbit	Sylvilagus floridanus
Short-tailed Shrew	Blarina brevicauda

PLANTS

Seaweeds:

Chenille Weeds	Dasya, several species
Coralline algae	Corallina officinalis
Dulse	Rhodymenia palmata
Enteromorpha	Enteromorpha intestinalis
Irish Moss	Chondrus crispus
Kelps	Laminaria, several species; Agarum cribrosum; Alaria esculenta
Knotted Wrack	Ascophyllum nodosum
Mermaid's Tresses	Cladophora gracilis
Rockweeds	Fucus, several species
Sargassum Weed	Sargassum vulgare
Sea Lettuce	Ulva lactuca

Lichens, Mosses and Fungi:

British Soldier Lichen	Cladonia cristatella
Earth Star	Geaster hygrometricus
Iceland Moss	Cetraria islandica
Old Man's Beard	Usnea barbata
Reindeer Moss	Cladonia tenuis, alpestris and rangiferina
Sphagnum Moss	Sphagnum, several species
Yellow Wall Lichen	Xanthoria parietina

Seed Plants:

Arrowhead	Sagittaria latifolia
Bayberry	Myrica pensylvanica
Beach Clotbur	Xanthium echinatum
Beach Grass	Ammophila breviligulata
Beach Heather	Hudsonia tomentosa
Beach Pea	Lathyrus maritimus
Beach Plum	Prunus maritima
Beach Wormwood	Artemisia stelleriana
Bearberry	Arctostaphylos uva-ursi
Black Grass	Juncus gerardi
Black Gum tree	Nyassa sylvatica

Blueberry	Vaccinium vacillans
Broom Crowberry	Corema conradii
Bulrush	Scirpus validus
Buttonbush	Cephalanthus occidentalis
Cattail	Typha latifolia
Cord Grass	Spartina alterniflora
Cranberry	Vaccinium macrocarpon
Duckweed	Spirodela polyrhiza
Eelgrass	Zostera marina
Golden Aster	Chrysopsis falcata
Golden Heather	Hudsonia ericoides
Hardhack	Spiraea tomentosa
Heather	Calluna vulgaris
Highbush Blueberry	Vaccinium corymbosum
Holly tree	Ilex opaca
Indian Pipe	Monotropa uniflora
Jointed Glasswort	Salicornia europaea
Meadowsweet	Spiraea latifolia
Moccasin Flower	Fissipes acaulis
Partridge Berry	Mitchella repens
Pickerel-weed	Pontederia cordata
Pink Azalea	Rhododendron nudiflorum
Pipewort	Eriocaulon articulatum
Pitch Pine	Pinus rigida
Pitcher Plant	Sarracenia purpurea
Plume Grass	Phragmites communis
Poison Ivy	Rhus radicans
Pond-lily	Nymphaea advena
Pondweed	Potamogeton natans
Prickly Pear	Opuntia vulgaris
Red Cedar	Juniperus virginiana
Rose Mallow	Hibiscus moscheutos
Rose Pogonia	Pogonia ophioglossoides
Salt-meadow Grass	Spartina patens
Salt-spray Rose	Rosa rugosa
Saltwort	Salsola kali
Sea Blite	Suaeda maritima
Sea Lavender	Limonium carolinianum
Sea Rocket	Cakile edentula
Sedges	Carex, several species
Seabeach Orache	Atriplex hastata
Seabeach Sandwort	Arenaria peploides
Seaside Aster	Aster linariifolius
Seaside Gerardia	Gerardia maritima
Seaside Goldenrod	Solidago sempervirens
Seaside Spurge	Euphorbia polyganifolia

Scrub Oak	Quercus ilicifolia
Sheep Laurel	Kalmia angustifolia
Spike Grass	Distichlis spicata
Sundew	Drosera rotundifolia
Sweet Pepper Bush	Clethra alnifolia
Tall Wormwood	Artemisia caudata
Water-lily	Castalia odorata
Waterweed	Elodea canadensis
Wild Cherry tree	Prunus serotina
Wild Indigo	Baptisia tinctoria
Woody Glasswort	Salicornia ambigua

INDEX

Figures in bold face indicate illustrations.

C

D

E

F

G

H

I

J

K

L

M

N

O

P

Q

R

S

T

U

V

W

X

Y